HOLT

Physics

Problem Workbook

HOLT, RINEHART AND WINSTON

A Harcourt Education Company

Orlando • **Austin** • New York • San Diego • Toronto • London

Printed in the United States of America

ISBN 0-03-036833-2

12 13 14 15 16 17 18 0956 17 16 15 14 13 12 11 10
4500230348

Contents

Problem	Title	Page

Magnetism

Electromagnetic Induction

Atomic Physics

Subatomic Physics

Advanced Topics

NAME ______________ DATE ______________ CLASS ______________

The Science of Physics

Problem A

METRIC PREFIXES

PROBLEM

In Hindu chronology, the longest time measure is a *para.* One *para* equals 311 040 000 000 000 years. Calculate this value in megahours and in nanoseconds. Write your answers in scientific notation.

SOLUTION

Given: 1 para = 311 040 000 000 000 years

Unknown: 1 para = ? Mh

1 para = ? ns

Express the time in years in terms of scientific notation. Then build conversion factors from the relationships given in **Table 3.**

$$1 \text{ para} = 3.1104 \times 10^{14} \text{ years}$$

$$\frac{365.25 \text{ days}}{1 \text{ year}} \times \frac{24 \text{ h}}{1 \text{ day}} \times \frac{1 \text{ Mh}}{1 \times 10^6 \text{ h}}$$

$$\frac{365.25 \text{ days}}{1 \text{ year}} \times \frac{24 \text{ h}}{1 \text{ day}} \times \frac{3600 \text{ s}}{1 \text{ h}} \times \frac{1 \text{ ns}}{1 \times 10^{-9} \text{s}}$$

Convert from years to megahours by multiplying the time by the first conversion expression.

$$1 \text{ para} = 3.1104 \times 10^{14} \cancel{\text{years}} \times \frac{365.25 \cancel{\text{days}}}{1 \cancel{\text{year}}} \times \frac{24 \cancel{\text{h}}}{1 \cancel{\text{day}}} \times \frac{1 \text{ Mh}}{1 \times 10^6 \cancel{\text{h}}}$$

$$= \boxed{2.7266 \times 10^{12} \text{ Mh}}$$

Convert from years to nanoseconds by multiplying the time by the second conversion expression.

$$1 \text{ para} = 3.1104 \times 10^{14} \cancel{\text{years}} \times \frac{365.25 \cancel{\text{days}}}{1 \cancel{\text{year}}} \times \frac{24 \cancel{\text{h}}}{1 \cancel{\text{day}}} \times \frac{3600 \cancel{\text{s}}}{1 \cancel{\text{h}}} \times \frac{1 \text{ ns}}{1 \times 10^{-9} \cancel{\text{s}}}$$

$$= \boxed{9.8157 \times 10^{30} \text{ ns}}$$

ADDITIONAL PRACTICE

1. One light-year is the distance light travels in one year. This distance is equal to 9.461×10^{15} m. After the sun, the star nearest to Earth is Alpha Centauri, which is about 4.35 light-years from Earth. Express this distance in

a. megameters.

b. picometers.

2. It is estimated that the sun will exhaust all of its energy in about ten billion years. By that time, it will have radiated about 1.2×10^{44} J (joules) of energy. Express this amount of energy in

a. kilojoules.
b. nanojoules.

3. The smallest living organism discovered so far is called a *mycoplasm.* Its mass is estimated as 1.0×10^{-16} g. Express this mass in

a. petagrams.
b. femtograms.
c. attograms.

4. The "extreme" prefixes that are officially recognized are *yocto,* which indicates a fraction equal to 10^{-24}, and *yotta,* which indicates a factor equal to 10^{24}. The maximum distance from Earth to the sun is 152 100 000 km. Using scientific notation, express this distance in

a. yoctometers (ym).
b. yottameters (Ym).

5. In 1993, the total production of nuclear energy in the world was 2.1×10^{15} watt-hours, where a watt is equal to one joule (J) per second. Express this number in

a. joules.
b. gigajoules.

6. In Einstein's special theory of relativity, mass and energy are equivalent. An expression of this equivalence can be made in terms of electron volts (units of energy) and kilograms, with one electron volt (eV) being equal to 1.78×10^{-36} kg. Using this ratio, express the mass of the heaviest mammal on earth, the blue whale, which has an average mass of 1.90×10^{5} kg, in

a. mega electron volts.
b. tera electron volts.

7. The most massive star yet discovered in our galaxy is one of the stars in the Carina Nebula, which can be seen from Earth's Southern Hemisphere and from the tropical latitudes of the Northern Hemisphere. The star, designated as Eta Carinae, is believed to be 200 times as massive as the sun, which has a mass of nearly 2×10^{30} kg. Find the mass of Eta Carinae in

a. milligrams.
b. exagrams.

8. The Pacific Ocean has a surface area of about 166 241 700 km^2 and an average depth of 3940 m. Estimate the volume of the Pacific Ocean in

a. cubic centimeters.
b. cubic millimeters.

Motion in One Dimension

Problem A

AVERAGE VELOCITY AND DISPLACEMENT

PROBLEM

The fastest fish, the sailfish, can swim 1.2×10^2 km/h. Suppose you have a friend who lives on an island 16 km away from the shore. If you send a message using a sailfish as a messenger, how long will it take for the message to reach your friend?

SOLUTION

Given: $v_{avg} = 1.2 \times 10^2$ km/h

$\Delta x = 16$ km

Unknown: $\Delta t = ?$

Use the definition of average speed to find Δt.

$$v_{avg} = \frac{\Delta x}{\Delta t}$$

Rearrange the equation to calculate Δt.

$$\Delta t = \frac{\Delta x}{v_{avg}}$$

$$\Delta t = \frac{16 \text{ km}}{\left(1.2 \times 10^2 \frac{\text{km}}{\text{h}}\right)\left(\frac{1 \text{ h}}{60 \text{ min}}\right)} = \frac{16 \text{ km}}{2.0 \text{ km/min}}$$

$$= \boxed{8.0 \text{ min}}$$

ADDITIONAL PRACTICE

1. The Sears Tower in Chicago is 443 m tall. Joe wants to set the world's stair climbing record and runs all the way to the roof of the tower. If Joe's average upward speed is 0.60 m/s, how long will it take Joe to climb from street level to the roof of the Sears Tower?

2. An ostrich can run at speeds of up to 72 km/h. How long will it take an ostrich to run 1.5 km at this top speed?

3. A cheetah is known to be the fastest mammal on Earth, at least for short runs. Cheetahs have been observed running a distance of 5.50×10^2 m with an average speed of 1.00×10^2 km/h.
   a. How long would it take a cheetah to cover this distance at this speed?
   b. Suppose the average speed of the cheetah were just 85.0 km/h. What distance would the cheetah cover during the same time interval calculated in (a)?

4. A pronghorn antelope has been observed to run with a top speed of 97 km/h. Suppose an antelope runs 1.5 km with an average speed of 85 km/h, and then runs 0.80 km with an average speed of 67 km/h.

a. How long will it take the antelope to run the entire 2.3 km?
b. What is the antelope's average speed during this time?

5. Jupiter, the largest planet in the solar system, has an equatorial radius of about 7.1×10^4 km (more than 10 times that of Earth). Its period of rotation, however, is only 9 h, 50 min. That means that every point on Jupiter's equator "goes around the planet" in that interval of time. Calculate the average speed (in m/s) of an equatorial point during one period of Jupiter's rotation. Is the average velocity different from the average speed in this case?

6. The peregrine falcon is the fastest of flying birds (and, as a matter of fact, is the fastest living creature). A falcon can fly 1.73 km downward in 25 s. What is the average velocity of a peregrine falcon?

7. The black mamba is one of the world's most poisonous snakes, and with a maximum speed of 18.0 km/h, it is also the fastest. Suppose a mamba waiting in a hide-out sees prey and begins slithering toward it with a velocity of +18.0 km/h. After 2.50 s, the mamba realizes that its prey can move faster than it can. The snake then turns around and slowly returns to its hide-out in 12.0 s. Calculate

a. the mamba's average velocity during its return to the hideout.
b. the mamba's average velocity for the complete trip.
c. the mamba's average speed for the complete trip.

8. In the Netherlands, there is an annual ice-skating race called the "Tour of the Eleven Towns." The total distance of the course is 2.00×10^2 km, and the record time for covering it is 5 h, 40 min, 37 s.

a. Calculate the average speed of the record race.
b. If the first half of the distance is covered by a skater moving with a speed of $1.05v$, where v is the average speed found in (a), how long will it take to skate the first half? Express your answer in hours and minutes.

NAME ______________________ DATE ______________ CLASS ______________

Motion in One Dimension

Problem B

AVERAGE ACCELERATION

PROBLEM

In 1977 off the coast of Australia, the fastest speed by a vessel on the water was achieved. If this vessel were to undergo an average acceleration of 1.80 m/s², it would go from rest to its top speed in 85.6 s. What was the speed of the vessel?

SOLUTION

Given: $a_{avg} = 1.80\ \text{m/s}^2$

$\Delta t = 85.6\ \text{s}$

$v_i = 0\ \text{m/s}$

Unknown: $v_f = ?$

Use the definition of average acceleration to find v_f.

$$a_{avg} = \frac{\Delta v}{\Delta t} = \frac{v_f - v_i}{\Delta t}$$

Rearrange the equation to calculate v_f.

$$v_f = a_{avg}\,\Delta t + v_i$$

$$v_f = \left(1.80\ \frac{\text{m}}{\text{s}^2}\right)\left(85.6\ \text{s}\right) + 0\ \frac{\text{m}}{\text{s}}$$

$$= 154\ \frac{\text{m}}{\text{s}}$$

$$= \left(154\ \frac{\text{m}}{\text{s}}\right)\left(\frac{3.60 \times 10^3\ \text{s}}{1\ \text{h}}\right)\left(\frac{1\ \text{km}}{10^3\ \text{m}}\right)$$

$$= \boxed{554\ \frac{\text{km}}{\text{h}}}$$

ADDITIONAL PRACTICE

1. If the vessel in the sample problem accelerates for 1.00 min, what will its speed be after that minute? Calculate the answer in both meters per second and kilometers per hour.

2. In 1935, a French destroyer, *La Terrible,* attained one of the fastest speeds for any standard warship. Suppose it took 2.0 min at a constant acceleration of 0.19 m/s² for the ship to reach its top speed after starting from rest. Calculate the ship's final speed.

3. In 1934, the wind speed on Mt. Washington in New Hampshire reached a record high. Suppose a very sturdy glider is launched in this wind, so that in 45.0 s the glider reaches the speed of the wind. If the

glider undergoes a constant acceleration of 2.29 m/s^2, what is the wind's speed? Assume that the glider is initially at rest.

4. In 1992, Maurizio Damilano, of Italy, walked 29 752 m in 2.00 h.

a. Calculate Damilano's average speed in m/s.

b. Suppose Damilano slows down to 3.00 m/s at the midpoint in his journey, but then picks up the pace and accelerates to the speed calculated in (a). It takes Damilano 30.0 s to accelerate. Find the magnitude of the average acceleration during this time interval.

5. South African frogs are capable of jumping as far as 10.0 m in one hop. Suppose one of these frogs makes exactly 15 of these jumps in a time interval of 60.0 s.

a. What is the frog's average velocity?

b. If the frog lands with a velocity equal to its average velocity and comes to a full stop 0.25 s later, what is the frog's average acceleration?

6. In 1991 at Smith College, in Massachusetts, Ferdie Adoboe ran 1.00×10^2 m backward in 13.6 s. Suppose it takes Adoboe 2.00 s to achieve a velocity equal to her average velocity during the run. Find her average acceleration during the first 2.00 s.

7. In the 1992 Summer Olympics, the German four-man kayak team covered 1 km in just under 3 minutes. Suppose that between the starting point and the 150 m mark the kayak steadily increases its speed from 0.0 m/s to 6.0 m/s, so that its average speed is 3.0 m/s.

a. How long does it take to cover the 150 m?

b. What is the magnitude of the average acceleration during that part of the course?

8. The highest speed ever achieved on a bicycle was reached by John Howard of the United States. The bicycle, which was accelerated by being towed by a vehicle, reached a velocity of +245 km/h. Suppose Howard wants to slow down, and applies the brakes on his now freely moving bicycle. If the average acceleration of the bicycle during braking is –3.0 m/s^2, how long will it take for the bicycle's velocity to decrease by 20.0 percent?

9. In 1993, bicyclist Rebecca Twigg of the United States traveled 3.00 km in 217.347 s. Suppose Twigg travels the entire distance at her average speed and that she then accelerates at –1.72 m/s^2 to come to a complete stop after crossing the finish line. How long does it take Twigg to come to a stop?

10. During the Winter Olympic games at Lillehammer, Norway, in 1994, Dan Jansen of the United States skated 5.00×10^2 m in 35.76 s. Suppose it takes Jansen 4.00 s to increase his velocity from zero to his maximum velocity, which is 10.0 percent greater than his average velocity during the whole run. Calculate Jansen's average acceleration during the first 4.00 s.

NAME ______________________ DATE ____________ CLASS ____________

Motion in One Dimension

Problem C

DISPLACEMENT WITH CONSTANT ACCELERATION

PROBLEM

In England, two men built a tiny motorcycle with a wheel base (the distance between the centers of the two wheels) of just 108 mm and a wheel's measuring 19 mm in diameter. The motorcycle was ridden over a distance of 1.00 m. Suppose the motorcycle has constant acceleration as it travels this distance, so that its final speed is 0.800 m/s. How long does it take the motorcycle to travel the distance of 1.00 m? Assume the motorcycle is initially at rest.

SOLUTION

Given: $v_f = 0.800$ m/s

$v_i = 0$ m/s

$\Delta x = 1.00$ m

Unknown: $\Delta t = ?$

Use the equation for displacement with constant acceleration.

$$\Delta x = \frac{1}{2}(v_i + v_f)\Delta t$$

Rearrange the equation to calculate Δt.

$$\Delta t = \frac{2\Delta x}{v_f + v_i}$$

$$\Delta t = \frac{(2)(1.00 \text{ m})}{0.800 \frac{\text{m}}{\text{s}} + 0 \frac{\text{m}}{\text{s}}} = \frac{2.00}{0.800} \text{ s}$$

$$= \boxed{2.50 \text{ s}}$$

ADDITIONAL PRACTICE

1. In 1993, Ileana Salvador of Italy walked 3.0 km in under 12.0 min. Suppose that during 115 m of her walk Salvador is observed to steadily increase her speed from 4.20 m/s to 5.00 m/s. How long does this increase in speed take?
2. In a scientific test conducted in Arizona, a special cannon called HARP (High Altitude Research Project) shot a projectile straight up to an altitude of 180.0 km. If the projectile's initial speed was 3.00 km/s, how long did it take the projectile to reach its maximum height?
3. The fastest speeds traveled on land have been achieved by rocket-powered cars. The current speed record for one of these vehicles is about 1090 km/h, which is only 160 km/h less than the speed of sound in air. Suppose a car that is capable of reaching a speed of

1.09×10^3 km/h is tested on a flat, hard surface that is 25.0 km long. The car starts at rest and just reaches a speed of 1.09×10^3 km/h when it passes the 20.0 km mark.

a. If the car's acceleration is constant, how long does it take to make the 20.0 km drive?

b. How long will it take the car to decelerate if it goes from its maximum speed to rest during the remaining 5.00 km stretch?

4. In 1990, Dave Campos of the United States rode a special motorcycle called the *Easyrider* at an average speed of 518 km/h. Suppose that at some point Campos steadily decreases his speed from 100.0 percent to 60.0 percent of his average speed during an interval of 2.00 min. What is the distance traveled during that time interval?

5. A German stuntman named Martin Blume performed a stunt called "the wall of death." To perform it, Blume rode his motorcycle for seven straight hours on the wall of a large vertical cylinder. His average speed was 45.0 km/h. Suppose that in a time interval of 30.0 s Blume increases his speed steadily from 30.0 km/h to 42.0 km/h while circling inside the cylindrical wall. How far does Blume travel in that time interval?

6. An automobile that set the world record for acceleration increased speed from rest to 96 km/h in 3.07 s. How far had the car traveled by the time the final speed was achieved?

7. In a car accident involving a sports car, skid marks as long as 290.0 m were left by the car as it decelerated to a complete stop. The police report cited the speed of the car before braking as being "in excess of 100 mph" (161 km/h). Suppose that it took 10.0 seconds for the car to stop. Estimate the speed of the car before the brakes were applied. (REMINDER: Answer should read, "speed in excess of . . .")

8. Col. Joe Kittinger of the United States Air Force crossed the Atlantic Ocean in nearly 86 hours. The distance he traveled was 5.7×10^3 km. Suppose Col. Kittinger is moving with a constant acceleration during most of his flight and that his final speed is 10.0 percent greater than his initial speed. Find the initial speed based on this data.

9. The polar bear is an excellent swimmer, and it spends a large part of its time in the water. Suppose a polar bear wants to swim from an ice floe to a particular point on shore where it knows that seals gather. The bear dives into the water and begins swimming with a speed of 2.60 m/s. By the time the bear arrives at the shore, its speed has decreased to 2.20 m/s. If the polar bear's swim takes exactly 9.00 min and it has a constant deceleration, what is the distance traveled by the polar bear?

Motion in One Dimension

Problem D

VELOCITY AND DISPLACEMENT WITH CONSTANT ACCELERATION

PROBLEM

Some cockroaches can run as fast as 1.5 m/s. Suppose that two cockroaches are separated by a distance of 60.0 cm and that they begin to run toward each other at the same moment. Both insects have constant acceleration until they meet. The first cockroach has an acceleration of $0.20\ m/s^2$ in one direction, and the second one has an acceleration of $0.12\ m/s^2$ in the opposite direction. How much time passes before the two insects bump into each other?

SOLUTION

1. DEFINE

Given:

$a_1 = 0.20\ m/s^2$ (first cockroach's acceleration)

$v_{i,1} = 0\ m/s$ (first cockroach's initial speed)

$a_2 = 0.12\ m/s^2$ (second cockroach's acceleration)

$v_{i,2} = 0\ m/s$ (second cockroach's initial speed)

$d = 60.0\ cm = 0.60\ m$ (initial distance between the insects)

Unknown: $\Delta x_1 = ?$ $\Delta x_2 = ?$ $\Delta t = ?$

2. PLAN

Choose an equation(s) or situation: Use the equation for displacement with constant acceleration for each cockroach.

$$\Delta x_1 = v_{i,1}\Delta t + \frac{1}{2}a_1\Delta t^2$$

$$\Delta x_2 = v_{i,2}\Delta t + \frac{1}{2}a_2\Delta t^2$$

The distance the second cockroach travels can be expressed as the difference between the total distance that initially separates the two insects and the distance that the first insect travels.

$$\Delta x_2 = d - \Delta x_1$$

Rearrange the equation(s) to isolate the unknown(s): Substitute the expression for the first cockroach's displacement into the equation for the second cockroach's displacement using the equation relating the two displacements to the initial distance between the insects.

$$\Delta x_2 = d - \Delta x_1 = d - \left(v_{i,1}\Delta t + \frac{1}{2}a_1\Delta t^2\right)$$

$$= v_{i,2}\Delta t + \frac{1}{2}a_2\Delta t^2$$

The equation can be rewritten to express Δt in terms of the known quantities. To simplify the calculation, the terms involving the initial speeds, which are both zero, can be removed from the equations.

$$d - \frac{1}{2}a_1\Delta t^2 = \frac{1}{2}a_2\Delta t^2$$

$$\Delta t = \sqrt{\frac{2d}{a_1 + a_2}}$$

3. CALCULATE **Substitute the values into the equation(s) and solve:**

$$\Delta t = \sqrt{\frac{(2)(0.60\ \text{m})}{0.20\frac{\text{m}}{\text{s}^2} + 0.12\frac{\text{m}}{\text{s}^2}}} = \sqrt{\frac{1.2\ \text{m}}{0.32\ \text{m/s}^2}} = \boxed{1.9\ \text{s}}$$

4. EVALUATE The final speeds for the first and second cockroaches are 0.38 m/s and 0.23 m/s, respectively. Both of these values are well below the maximum speed for cockroaches in general.

ADDITIONAL PRACTICE

1. In 1986, the first flight around the globe without a single refueling was completed. The aircraft's average speed was 186 km/h. If the airplane landed at this speed and accelerated at $-1.5\ \text{m/s}^2$, how long did it take for the airplane to stop?
2. In 1976, Gerald Hoagland drove a car over 8.0×10^2 km in reverse. Fortunately for Hoagland and motorists in general, the event took place on a special track. During this drive, Hoagland's average velocity was about -15.0 m/s. Suppose Hoagland decides during his drive to go forward. He applies the brakes, stops, and then accelerates until he moves forward at same speed he had when he was moving backward. How long would the entire reversal process take if the average acceleration during this process is $+2.5\ \text{m/s}^2$?
3. The first permanent public railway was built by George Stephenson and opened in Cleveland, Ohio, in 1825. The average speed of the trains was 24.0 km/h. Suppose a train moving at this speed accelerates $-0.20\ \text{m/s}^2$ until it reaches a speed of 8.0 km/h. How long does it take the train to undergo this change in speed?
4. The winding cages in mine shafts are used to move workers in and out of the mines. These cages move much faster than any commercial elevators. In one South African mine, speeds of up to 65.0 km/h are attained. The mine has a depth of 2072 m. Suppose two cages start their downward journey at the same moment. The first cage quickly attains the maximum speed (an unrealistic situation), then proceeds to descend uniformly at that speed all the way to the bottom. The second cage starts at rest and then increases its speed with a constant acceleration of magnitude $4.00 \times 10^{-2}\ \text{m/s}^2$. How long will the trip take for each cage? Which cage will reach the bottom of the mine shaft first?
5. In a 1986 bicycle race, Fred Markham rode his bicycle a distance of 2.00×10^2 m with an average speed of 105.4 km/h. Markham and the bicycle started the race with a certain initial speed.
   - **a.** Find the time it took Markham to cover 2.00×10^2 m.
   - **b.** Suppose a car moves from rest under constant acceleration. What is the magnitude of the car's acceleration if the car is to finish the race at exactly the same time Markham finishes the race?

NAME ______________________ DATE ____________ CLASS ______________

6. Some tropical butterflies can reach speeds of up to 11 m/s. Suppose a butterfly flies at a speed of 6.0 m/s while another flying insect some distance ahead flies in the same direction with a constant speed. The butterfly then increases its speed at a constant rate of 1.4 m/s^2 and catches up to the other insect 3.0 s later. How far does the butterfly travel during the race?

7. Mary Rife, of Texas, set a women's world speed record for sailing. In 1977, her vessel, *Proud Mary,* reached a speed of 3.17×10^2 km/h. Suppose it takes 8.0 s for the boat to decelerate from 3.17×10^2 km/h to 2.00×10^2 km/h. What is the boat's acceleration? What is the displacement of the *Proud Mary* as it slows down?

8. In 1994, a human-powered submarine was designed in Boca Raton, Florida. It achieved a maximum speed of 3.06 m/s. Suppose this submarine starts from rest and accelerates at 0.800 m/s^2 until it reaches maximum speed. The submarine then travels at constant speed for another 5.00 s. Calculate the total distance traveled by the submarine.

9. The highest speed achieved by a standard nonracing sports car is 3.50×10^2 km/h. Assuming that the car accelerates at 4.00 m/s^2, how long would this car take to reach its maximum speed if it is initially at rest? What distance would the car travel during this time?

10. Stretching 9345 km from Moscow to Vladivostok, the Trans-Siberian railway is the longest single railroad in the world. Suppose the train is approaching the Moscow station at a velocity of +24.7 m/s when it begins a constant acceleration of –0.850 m/s^2. This acceleration continues for 28 s. What will be the train's final velocity when it reaches the station?

11. The world's fastest warship belongs to the United States Navy. This vessel, which floats on a cushion of air, can move as fast as 1.7×10^2 km/h. Suppose that during a training exercise the ship accelerates +2.67 m/s^2, so that after 15.0 s its displacement is $+6.00 \times 10^2$ m. Calculate the ship's initial velocity just before the acceleration. Assume that the ship moves in a straight line.

12. The first supersonic flight was performed by then Capt. Charles Yeager in 1947. He flew at a speed of 3.00×10^2 m/s at an altitude of more than 12 km, where the speed of sound in air is slightly less than 3.00×10^2 m/s. Suppose Capt. Yeager accelerated 7.20 m/s^2 in 25.0 s to reach a final speed of 3.00×10^2 m/s. What was his initial speed?

13. Peter Rosendahl rode his unicycle a distance of 1.00×10^2 m in 12.11 s. If Rosendahl started at rest, what was the magnitude of his acceleration?

14. Suppose that Peter Rosendahl began riding the unicycle with a speed of 3.00 m/s and traveled a distance of 1.00×10^2 m in 12.11s. What would the magnitude of Rosendahl's acceleration be in this case?

15. In 1991, four English teenagers built an electric car that could attain a speed 30.0 m/s. Suppose it takes 8.0 s for this car to accelerate from 18.0 m/s to 30.0 m/s. What is the magnitude of the car's acceleration?

NAME ______________________ DATE ______________ CLASS ______________

Motion in One Dimension

Problem E

FINAL VELOCITY AFTER ANY DISPLACEMENT

PROBLEM

In 1970, a rocket-powered car called *Blue Flame* achieved a maximum speed of 1.00 (10^3 km/h (278 m/s). Suppose the magnitude of the car's constant acceleration is 5.56 m/s^2. If the car is initially at rest, what is the distance traveled during its acceleration?

SOLUTION

1. DEFINE

Given:

$v_i = 0$ m/s

$v_f = 278$ m/s

$a = 5.56$ m/s^2

Unknown: $\Delta x = ?$

2. PLAN

Choose an equation(s) or situation: Use the equation for the final velocity after any displacement.

$$v_f^2 = v_i^2 + 2a\Delta x$$

Rearrange the equation(s) to isolate the unknown(s):

$$\Delta x = \frac{v_f^2 - v_i^2}{2a}$$

3. CALCULATE

Substitute the values into the equation(s) and solve:

$$\Delta x = \frac{\left(278 \frac{\text{m}}{\text{s}}\right)^2 - \left(0 \frac{\text{m}}{\text{s}}\right)^2}{(2)\left(5.56 \frac{\text{m}}{\text{s}^2}\right)} = \boxed{6.95 \times 10^3 \text{ m}}$$

4. EVALUATE

Using the appropriate kinematic equation, the time of travel for *Blue Flame* is found to be 50.0 s. From this value for time the distance traveled during the acceleration is confirmed to be almost 7 km. Once the car reaches its maximum speed, it travels about 16.7 km/min.

ADDITIONAL PRACTICE

1. In 1976, Kitty Hambleton of the United States drove a rocket-engine car to a maximum speed of 965 km/h. Suppose Kitty started at rest and underwent a constant acceleration with a magnitude of 4.0 m/s^2. What distance would she have had to travel in order to reach the maximum speed?

2. With a cruising speed of 2.30×10^3 km/h, the French supersonic passenger jet Concorde is the fastest commercial airplane. Suppose the landing speed of the Concorde is 20.0 percent of the cruising speed. If the plane accelerates at -5.80 m/s^2, how far does it travel between the time it lands and the time it comes to a complete stop?

3. The Boeing 747 can carry more than 560 passengers and has a maximum speed of about 9.70×10^2 km/h. After takeoff, the plane takes a certain time to reach its maximum speed. Suppose the plane has a constant acceleration with a magnitude of 4.8 m/s^2. What distance does the plane travel between the moment its speed is 50.0 percent of maximum and the moment its maximum speed is attained?

4. The distance record for someone riding a motorcycle on its rear wheel without stopping is more than 320 km. Suppose the rider in this unusual situation travels with an initial speed of 8.0 m/s before speeding up. The rider then travels 40.0 m at a constant acceleration of 2.00 m/s^2. What is the rider's speed after the acceleration?

5. The skid marks left by the decelerating jet-powered car *The Spirit of America* were 9.60 km long. If the car's acceleration was –2.00 m/s^2, what was the car's initial velocity?

6. The heaviest edible mushroom ever found (the so-called "chicken of the woods") had a mass of 45.4 kg. Suppose such a mushroom is attached to a rope and pulled horizontally along a smooth stretch of ground, so that it undergoes a constant acceleration of +0.35 m/s^2. If the mushroom is initially at rest, what will its velocity be after it has been displaced +64 m?

7. Bengt Norberg of Sweden drove his car 44.8 km in 60.0 min. The feature of this drive that is interesting is that he drove the car on two side wheels.

- **a.** Calculate the car's average speed.
- **b.** Suppose Norberg is moving forward at the speed calculated in (a). He then accelerates at a rate of –2.00 m/s^2. After traveling 20.0 m, the car falls on all four wheels. What is the car's final speed while still traveling on two wheels?

8. Starting at a certain speed, a bicyclist travels 2.00×10^2 m. Suppose the bicyclist undergoes a constant acceleration of 1.20 m/s^2. If the final speed is 25.0 m/s, what was the bicyclist's initial speed?

9. In 1994, Tony Lang of the United States rode his motorcycle a short distance of 4.0×10^2 m in the short interval of 11.5 s. He started from rest and crossed the finish line with a speed of about 2.50×10^2 km/h. Find the magnitude of Lang's acceleration as he traveled the 4.0×10^2 m distance.

10. The lightest car in the world was built in London and had a mass of less than 10 kg. Its maximum speed was 25.0 km/h. Suppose the driver of this vehicle applies the brakes while the car is moving at its maximum speed. The car stops after traveling 16.0 m. Calculate the car's acceleration.

NAME ______________________ DATE ______________ CLASS ______________

Motion in One Dimension

Problem F

FALLING OBJECT

PROBLEM

The famous Gateway to the West Arch in St. Louis, Missouri, is about 192 m tall at its highest point. Suppose Sally, a stuntwoman, jumps off the top of the arch. If it takes Sally 6.4 s to land on the safety pad at the base of the arch, what is her average acceleration? What is her final velocity?

SOLUTION

1. DEFINE

Given: $v_i = 0$ m/s

$\Delta y = -192$ m

$\Delta t = 6.4$ s

Unknown: $a = ?$

$v_f = ?$

2. PLAN

Choose an equation(s) or situation: Both the acceleration and the final speed are unknown. Therefore, first solve for the acceleration during the fall using the equation that requires only the known variables.

$$\Delta y = v_i\,\Delta t + \frac{1}{2}a\,\Delta t^2$$

Then the equation for v_f that involves acceleration can be used to solve for v_f.

$$v_f = v_i + a\Delta t$$

Rearrange the equation(s) to isolate the unknown(s):

$$a = \frac{2(\Delta y - v_i\,\Delta t)}{\Delta t^2}$$

$$v_f = v_i + a\Delta t$$

3. CALCULATE

Substitute the values into the equation(s) and solve:

$$a = \frac{(2)\left[(-192\text{ m}) - \left(0\frac{\text{m}}{\text{s}}\right)(6.4\text{s})\right]}{(6.4\text{ s})^2} = \boxed{-9.4\frac{\text{m}}{\text{s}^2}}$$

$$v_f = 0\frac{\text{m}}{\text{s}} + \left(-9.4\frac{\text{m}}{\text{s}^2}\right)(6.4\text{s}) = \boxed{-6.0\times 10^1\frac{\text{m}}{\text{s}}}$$

4. EVALUATE

Sally's downward acceleration is less than the free-fall acceleration at Earth's surface (9.81 m/s^2). This indicates that air resistance reduces her downward acceleration by 0.4 m/s^2. Sally's final speed, 60 m\s^2, is such that, if she could fall at this speed at the beginning of her jump with no acceleration, she would travel a distance equal to the arch's height in just a little more than 3 s.

NAME ______________________ DATE ____________ CLASS ____________

ADDITIONAL PRACTICE

1. The John Hancock Center in Chicago is the tallest building in the United States in which there are residential apartments. The Hancock Center is 343 m tall. Suppose a resident accidentally causes a chunk of ice to fall from the roof. What would bé the velocity of the ice as it hits the ground? Neglect air resistance.

2. Brian Berg of Iowa built a house of cards 4.88 m tall. Suppose Berg throws a ball from ground level with a velocity of 9.98 m/s straight up. What is the velocity of the ball as it first passes the top of the card house?

3. The Sears Tower in Chicago is 443 m tall. Suppose a book is dropped from the top of the building. What would be the book's velocity at a point 221 m above the ground? Neglect air resistance.

4. The tallest roller coaster in the world is the Desperado in Nevada. It has a lift height of 64 m. If an archer shoots an arrow straight up in the air and the arrow passes the top of the roller coaster 3.0 s after the arrow is shot, what is the initial speed of the arrow?

5. The tallest *Sequoia sempervirens* tree in California's Redwood National Park is 111 m tall. Suppose an object is thrown downward from the top of that tree with a certain initial velocity. If the object reaches the ground in 3.80 s, what is the object's initial velocity?

6. The Westin Stamford Hotel in Detroit is 228 m tall. If a worker on the roof drops a sandwich, how long does it take the sandwich to hit the ground, assuming there is no air resistance? How would air resistance affect the answer?

7. A man named Bungkas climbed a palm tree in 1970 and built himself a nest there. In 1994 he was still up there, and he had not left the tree for 24 years. Suppose Bungkas asks a villager for a newspaper, which is thrown to him straight up with an initial speed of 12.0 m/s. When Bungkas catches the newspaper from his nest, the newspaper's velocity is 3.0 m/s, directed upward. From this information, find the height at which the nest was built. Assume that the newspaper is thrown from a height of 1.50 m above the ground.

8. Rob Colley set a record in "pole-sitting" when he spent 42 days in a barrel at the top of a flagpole with a height of 43 m. Suppose a friend wanting to deliver an ice-cream sandwich to Colley throws the ice cream straight up with just enough speed to reach the barrel. How long does it take the ice-cream sandwich to reach the barrel?

9. A common flea is recorded to have jumped as high as 21 cm. Assuming that the jump is entirely in the vertical direction and that air resistance is insignificant, calculate the time it takes the flea to reach a height of 7.0 cm.

NAME ______________________ DATE ____________ CLASS ______________

Two-Dimensional Motion and Vectors

Problem A

FINDING RESULTANT MAGNITUDE AND DIRECTION

PROBLEM

Cheetahs are, for short distances, the fastest land animals. In the course of a chase, cheetahs can also change direction very quickly. Suppose a cheetah runs straight north for 5.0 s, quickly turns, and runs 3.00×10^2 m west. If the magnitude of the cheetah's resultant displacement is 3.35×10^2 m, what is the cheetah's displacement and velocity during the first part of its run?

SOLUTION

1. DEFINE **Given:**

$\Delta t_1 = 5.0$ s

$\Delta x = 3.00 \text{ s} \times 10^2$ m

$d = 3.35 \times 10^2$ m

Unknown: $\Delta y = ?$ $\quad v_y = ?$

Diagram:

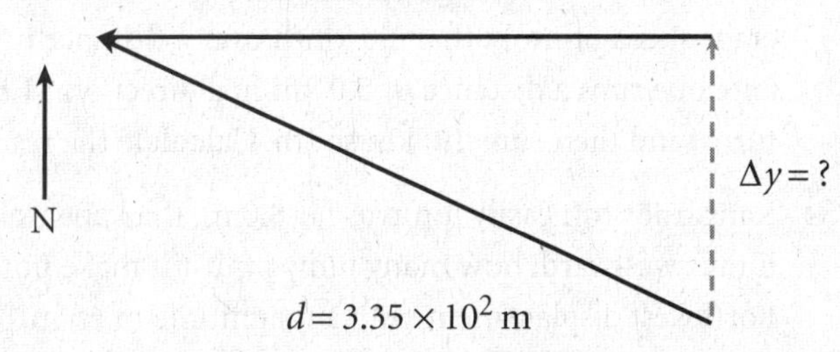

2. PLAN **Choose the equation(s) or situation:** Use the Pythagorean theorem to subtract one of the displacements at right angles from the total displacement, and thus determine the unknown component of displacement.

$$d^2 = \Delta x^2 + \Delta y^2$$

Use the equation relating displacement to constant velocity and time, and use the calculated value for Δy and the given value for Δt to solve for v.

$$\Delta v = \frac{\Delta y}{\Delta t}$$

Rearrange the equation(s) to isolate the unknown(s):

$$\Delta y^2 = d^2 - \Delta x^2$$

$$\Delta y = \sqrt{d^2 - \Delta x^2}$$

$$v_y = \frac{\Delta y}{\Delta t}$$

3. CALCULATE **Substitute the values into the equation(s) and solve:** Because the value for Δy is a displacement magnitude, only the positive root is used ($\Delta y > 0$).

$$\Delta y = \sqrt{(3.35 \times 10^2 \text{ m})^2 - (3.00 \times 10^2 \text{ m})^2}$$

$$= \sqrt{1.12 \times 10^5 \text{ m}^2 - 9.00 \times 10^4 \text{ m}^2}$$

$$= \sqrt{2.2 \times 10^4 \text{ m}}$$

$$= \boxed{1.5 \times 10^2 \text{ m, north}}$$

$$v_y = \frac{1.5 \times 10^2 \text{ m}}{5.0 \text{ s}} = \boxed{3.0 \times 10^1 \text{ m/s, north}}$$

4. EVALUATE The cheetah has a top speed of 30 m/s, or 107 km/h. This is equal to about 67 miles/h.

ADDITIONAL PRACTICE

1. An ostrich cannot fly, but it is able to run fast. Suppose an ostrich runs east for 7.95 s and then runs 161 m south, so that the magnitude of the ostrich's resultant displacement is 226 m. Calculate the magnitude of the ostrich's eastward component and its running speed.

2. The pronghorn antelope, found in North America, is the best long-distance runner among mammals. It has been observed to travel at an average speed of more than 55 km/h over a distance of 6.0 km. Suppose the antelope runs a distance of 5.0 km in a direction 11.5° north of east, turns, and then runs 1.0 km south. Calculate the resultant displacement.

3. Kangaroos can easily jump as far 8.0 m. If a kangaroo makes five such jumps westward, how many jumps must it make northward to have a northwest displacement with a magnitude of 68 m? What is the angle of the resultant displacement with respect to north?

4. In 1926, Gertrude Ederle of the United States became the first woman to swim across the English channel. Suppose Ederle swam 25.2 km east from the coast near Dover, England, then made a 90° turn and traveled south for 21.3 km to a point east of Calais, France. What was Ederle's resultant displacement?

5. The emperor penguin is the best diver among birds: the record dive is 483 m. Suppose an emperor penguin dives vertically to a depth of 483 m and then swims horizontally a distance of 225 m. What angle would the vector of the resultant displacement make with the water's surface? What is the magnitude of the penguin's resultant displacement?

6. A killer whale can swim as fast as 15 m/s. Suppose a killer whale swims in one direction at this speed for 8.0 s, makes a 90° turn, and continues swimming in the new direction with the same speed as before. After a certain time interval, the magnitude of the resultant displacement is 180.0 m. Calculate the amount of time the whale swims after changing direction.

7. Woodcocks are the slowest birds: their average speed during courtship displays can be as low as 8.00 km/h. Suppose a woodcock flies east for 15.0 min. It then turns and flies north for 22.0 min. Calculate the magnitude of the resultant displacement and the angle between the resultant displacement and the woodcock's initial displacement.

NAME ______________________ DATE ______________ CLASS ______________

Two-Dimensional Motion and Vectors

Problem B

RESOLVING VECTORS

PROBLEM

Certain iguanas have been observed to run as fast as 10.0 m/s. Suppose an iguana runs in a straight line at this speed for 5.00 s. The direction of motion makes an angle of 30.0° to the east of north. Find the value of the iguana's northward displacement.

SOLUTION

1. DEFINE

Given: $v = 10.0$ m/s
$t = 5.00$ s
$\theta = 30.0°$

Unknown: $\Delta y = ?$

Diagram:

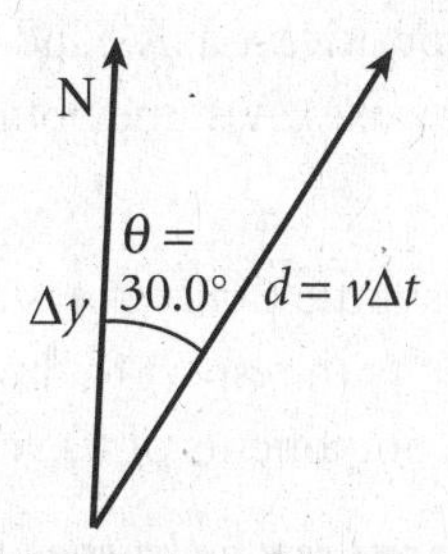

2. PLAN

Choose the equation(s) or situation: The northern component of the vector is equal to the vector magnitude times the cosine of the angle between the vector and the northward direction.

$$\Delta y = d(\cos \theta)$$

Use the equation relating displacement with constant velocity and time, and substitute it for d in the previous equation.

$$d = v\Delta t$$
$$\Delta y = v\Delta t(\cos \theta)$$

3. CALCULATE

Substitute the values into the equation(s) and solve:

$$\Delta y = \left(10.0 \frac{\text{m}}{\text{s}}\right)(5.00 \text{ s})(\cos 30.0°)$$

$$= \boxed{43.3 \text{ m, north}}$$

4. EVALUATE

The northern component of the displacement vector is smaller than the displacement itself, as expected.

ADDITIONAL PRACTICE

1. A common flea can jump a distance of 33 cm. Suppose a flea makes five jumps of this length in the northwest direction. If the flea's northward displacement is 88 cm, what is the flea's westward displacement.

2. The longest snake ever found was a python that was 10.0 m long. Suppose a coordinate system large enough to measure the python's length is drawn on the ground. The snake's tail is then placed at the origin and the snake's body is stretched so that it makes an angle of 60.0° with the positive *x*-axis. Find the *x* and *y* coordinates of the snake's head. (Hint: The *y*-coordinate is positive.)

3. A South-African sharp-nosed frog set a record for a triple jump by traveling a distance of 10.3 m. Suppose the frog starts from the origin of a coordinate system and lands at a point whose coordinate on the *y*-axis is equal to –6.10 m. What angle does the vector of displacement make with the negative *y*-axis? Calculate the *x* component of the frog.

4. The largest variety of grasshopper in the world is found in Malaysia. These grasshoppers can measure almost a foot (0.305 m) in length and can jump 4.5 m. Suppose one of these grasshoppers starts at the origin of a coordinate system and makes exactly eight jumps in a straight line that makes an angle of 35° with the positive *x*-axis. Find the grasshopper's displacements along the *x*- and *y*-axes. Assume both component displacements to be positive.

5. The landing speed of the space shuttle *Columbia* is 347 km/h. If the shuttle is landing at an angle of 15.0° with respect to the horizontal, what are the horizontal and the vertical components of its velocity?

6. In Virginia during 1994 Elmer Trett reached a speed of 372 km/h on his motorcycle. Suppose Trett rode northwest at this speed for 8.7 s. If the angle between east and the direction of Trett's ride was 60.0°, what was Trett's displacement east? What was his displacement north?

7. The longest delivery flight ever made by a twin-engine commercial jet took place in 1990. The plane covered a total distance of 14 890 km from Seattle, Washington to Nairobi, Kenya in 18.5 h. Assuming that the plane flew in a straight line between the two cities, find the magnitude of the average velocity of the plane. Also, find the eastward and southward components of the average velocity if the direction of the plane's flight was at an angle of 25.0° south of east.

8. The French bomber *Mirage IV* can fly over 2.3×10^3 km/h. Suppose this plane accelerates at a rate that allows it to increase its speed from 6.0×10^2 km/h to 2.3×10^3 km/h. in a time interval of 120 s. If this acceleration is upward and at an angle of 35° with the horizontal, find the acceleration's horizontal and vertical components.

NAME ______________ DATE ______________ CLASS ______________

Two-Dimensional Motion and Vectors

Problem C

ADDING VECTORS ALGEBRAICALLY

PROBLEM

The record for the longest nonstop closed-circuit flight by a model airplane was set in Italy in 1986. The plane flew a total distance of 1239 km. Assume that at some point the plane traveled 1.25×10^3 m to the east, then 1.25×10^3 m to the north, and finally 1.00×10^3 m to the southeast. Calculate the total displacement for this portion of the flight.

SOLUTION

1. DEFINE

Given: $d_1 = 1.25 \times 10^3$ m $\quad d_2 = 1.25 \times 10^3$ m $\quad d_3 = 1.00 \times 10^3$ m

Unknown: $\Delta x_{tot} = ?$ $\Delta y_{tot} = ?$ $\quad d = ?$ $\quad \theta = ?$

Diagram:

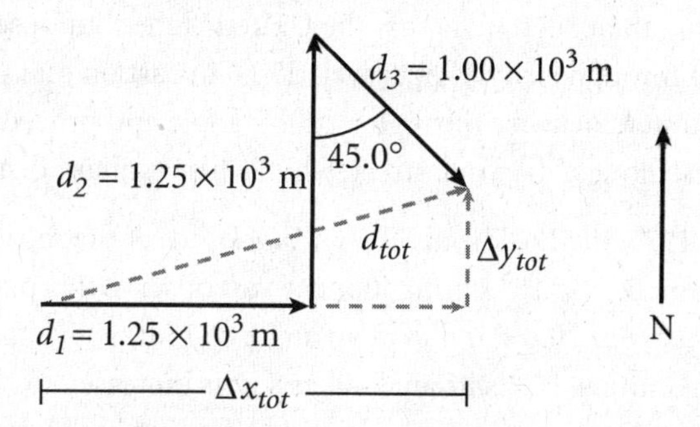

2. PLAN

Choose the equation(s) or situation: Orient the displacements with respect to the x-axis of the coordinate system.

$$\theta_1 = 0.00° \qquad \theta_2 = 90.0° \qquad \theta_3 = -45.0°$$

Use this information to calculate the components of the total displacement along the x-axis and the y-axis.

$$\Delta x_{tot} = \Delta x_1 + \Delta x_2 + \Delta x_3$$
$$= d_1(\cos \theta_1) + d_2(\cos \theta_2) + d_3(\cos \theta_3)$$
$$\Delta y_{tot} = \Delta y_1 + \Delta y_2 + \Delta y_3$$
$$= d_1(\sin \theta_1) + d_2(\sin \theta_2) + d_3(\sin \theta_3)$$

Use the components of the total displacement, the Pythagorean theorem, and the tangent function to calculate the total displacement.

$$d = \sqrt{(\Delta x_{tot})^2 + (\Delta y_{tot})^2} \qquad \theta = tan^{-1}\left(\frac{\Delta y_{tot}}{\Delta x_{tot}}\right)$$

3. CALCULATE

Substitute the values into the equation(s) and solve:

$$\Delta x_{tot} = (1.25 \times 10^3 \text{ m})(\cos 0°) + (1.25 \times 10^3 \text{ m})(\cos 90.0°)$$
$$+(1.00 \times 10^3 \text{ m})[\cos (-45.0°)]$$
$$= 1.25 \times 10^3 \text{ m} + 7.07 \times 10^2 \text{ m}$$
$$= 1.96 \times 10^3 \text{ m}$$
$$\Delta y_{tot} = (1.25 \times 10^3 \text{ m})(\sin 0°) + (1.25 \times 10^3 \text{ m})(\sin 90.0°)$$
$$+(1.00 \times 10^3 \text{ m})[\sin (-45.0°)]$$
$$= 1.25 \times 10^3 \text{ m} + 7.07 \times 10^2 \text{ m}$$
$$= 0.543 \times 10^3 \text{ m}$$

$$d = \sqrt{(1.96 \times 10^3 \text{ m})^2 + (0.543 \times 10^3 \text{ m})^2}$$

$$d = \sqrt{3.84 \times 10^6 \text{ m}^2 + 2.95 \times 10^5 \text{ m}^2} = \sqrt{4.14 \times 10^6 \text{ m}^2}$$

$$d = \boxed{2.03 \times 10^3 \text{ m}}$$

$$\theta = tan^{-1}\left(\frac{0.543 \times 10^3 \text{ m}}{1.96 \times 10^3 \text{ m}}\right)$$

$$\theta = \boxed{15.5° \text{ north of east}}$$

4. EVALUATE The magnitude of the total displacement is slightly larger than that of the total displacement in the eastern direction alone.

ADDITIONAL PRACTICE

1. For six weeks in 1992, Akira Matsushima, from Japan, rode a unicycle more than 3000 mi across the United States. Suppose Matsushima is riding through a city. If he travels 250.0 m east on one street, then turns counterclockwise through a 120.0° angle and proceeds 125.0 m northwest along a diagonal street, what is his resultant displacement?

2. In 1976, the Lockheed SR-71A *Blackbird* set the record speed for any airplane: 3.53×10^3 km/h. Suppose you observe this plane ascending at this speed. For 20.0 s, it flies at an angle of 15.0° above the horizontal, then for another 10.0 s its angle of ascent is increased to 35.0°. Calculate the plane's total gain in altitude, its total horizontal displacement, and its resultant displacement.

3. Magnor Mydland of Norway constructed a motorcycle with a wheelbase of about 12 cm. The tiny vehicle could be ridden at a maximum speed 11.6 km/h. Suppose this motorcycle travels in the directions d_1 and d_2 shown in the figure below. Calculate d_1 and d_2, and determine how long it takes the motorcycle to reach a net displacement of 2.0×10^2 m to the right?

4. The fastest propeller-driven aircraft is the Russian TU-95/142, which can reach a maximum speed of 925 km/h. For this speed, calculate the plane's resultant displacement if it travels east for 1.50 h, then turns 135° northwest and travels for 2.00 h.

5. In 1952, the ocean liner *United States* crossed the Atlantic Ocean in less than four days, setting the world record for commercial ocean-going vessels. The average speed for the trip was 57.2 km/h. Suppose the ship moves in a straight line eastward at this speed for 2.50 h. Then, due to a strong local current, the ship's course begins to deviate northward by 30.0°, and the ship follows the new course at the same speed for another 1.50 h. Find the resultant displacement for the 4.00 h period.

NAME ______________________ DATE ______________ CLASS ______________

Two-Dimensional Motion and Vectors

Problem D

PROJECTILES LAUNCHED HORIZONTALLY

PROBLEM

A movie director is shooting a scene that involves dropping s stunt dummy out of an airplane and into a swimming pool. The plane is 10.0 m above the ground, traveling at a velocity of 22.5 m/s in the positive *x* direction. The director wants to know where in the plane's path the dummy should be dropped so that it will land in the pool. What is the dummy's horizontal displacement?

SOLUTION

1. DEFINE

Given: $\Delta y = -10.0\ \mathrm{m}$ $\quad a_y = -g = -9.81\ \mathrm{m/s^2}$ $\quad v_x = 22.5\ \mathrm{m/s}$

Unknown: $\Delta t = ?$ $\quad \Delta x = ?$

Diagram: The initial velocity vector of the stunt dummy only has a horizontal component. Choose the coordinate system oriented so that the positive *y* direction points upward and the positive *x* direction points to the right.

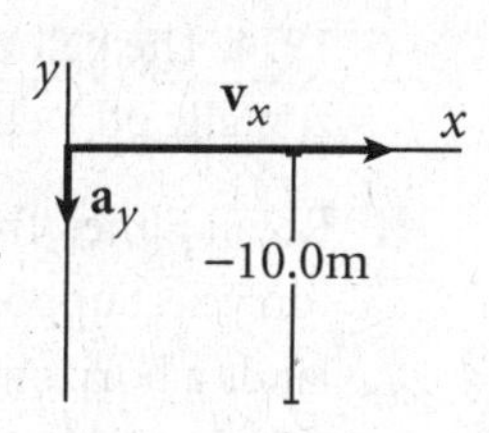

2. PLAN

Choose the equation(s) or situation: The dummy drops with no initial vertical velocity. Because air resistance is neglected, the dummy's horizontal velocity remains constant.

$$\Delta y = \tfrac{1}{2} a_y (\Delta t)^2$$

$$\Delta x = v_x \Delta t$$

Rearrange the equation(s) to isolate the unknown(s):

$$\Delta t = \sqrt{\frac{2\Delta y}{a_y}}$$

3. CALCULATE

First find the time it takes for the dummy to reach the ground.

$$\Delta t = \sqrt{\frac{(2)(-10.0\mathrm{m})}{(-9.81\ \mathrm{m/s^2})}} = 1.43\ \mathrm{s}$$

Find out how far horizontally the dummy can travel during this period of time.

$$\Delta x = v_x \Delta t = (22.5\ \mathrm{m/s})(1.43\ \mathrm{s})$$

$$= \boxed{32.2\ \mathrm{m}}$$

4. EVALUATE

The stunt dummy will have to drop from the plane when the plane is at a horizontal distance of 32.2 m from the pool. The distance is within the correct order of magnitude, given the other values in this problem.

NAME ______________________ DATE ____________ CLASS ____________

ADDITIONAL PRACTICE

1. Florence Griffith-Joyner of the United States set the women's world record for the 200 m run by running with an average speed of 9.37 m/s. Suppose Griffith-Joyner wants to jump over a river. She runs horizontally from the river's higher bank at 9.37 m/s and lands on the edge of the opposite bank. If the difference in height between the two banks is 2.00 m, how wide is the river?

2. The longest banana split ever made was 7.320 km long (needless to say, more than one banana was used). If an archer were to shoot an arrow horizontally from the top of Mount Everest, which is located 8848 m above sea level, would the arrow's horizontal displacement be larger than 7.32 km? Assume that the arrow cannot be shot faster than 100.0 m/s, that there is no air resistance, and that the arrow lands at sea level.

3. The longest shot on a golf tournament was made by Mike Austin in 1974. The ball went a distance of 471 m. Suppose the ball was shot horizontally off a cliff at 80.0 m/s. Calculate the height of the cliff.

4. Recall Elmer Trett, who in 1994 reached a speed of 372 km/h on his motorcycle. Suppose Trett drives off a horizontal ramp at this speed and lands a horizontal distance of 40.0 m away from the edge of the ramp. What is the height of the ramp? Neglect air resistance.

5. A Snorkel fire engine is designed for putting out fires that are well above street level. The engine has a hydraulic lift that lifts the firefighter and a system that delivers pressurized water to the firefighter. Suppose that the engine cannot move closer than 25 m to a building that has a fire on its sixth floor, which is 25 m above street level. Also assume that the water nozzle is stuck in the horizontal position (an improbable situation). If the horizontal speed of the water emerging from the hose is 15 m/s, how high above the street must the firefighter be lifted in order for the water to reach the fire?

6. The longest stuffed toy ever manufactured is a 420 m snake made by Norwegian children. Suppose a projectile is thrown horizontally from a height half as long as the snake and the projectile's horizontal displacement is as long as the snake. What would be the projectile's initial speed?

7. Libyan basketball player Suleiman Nashnush was the tallest basketball player ever. His height was 2.45 m. Suppose Nashnush throws a basketball horizontally from a level equal to the top of his head. If the speed of the basketball is 12.0 m/s when it lands, what was the ball's initial speed? (Hint: Consider the components of final velocity.)

8. The world's largest flowerpot is 1.95 m high. If you were to jump horizontally from the top edge of this flowerpot at a speed of 3.0 m/s, what would your landing velocity be?

Two-Dimensional Motion and Vectors

Problem E

PROJECTILES LAUNCHED AT AN ANGLE

PROBLEM

The narrowest strait on earth is Seil Sound in Scotland, which lies between the mainland and the island of Seil. The strait is only about 6.0 m wide. Suppose an athlete wanting to jump "over the sea" leaps at an angle of 35° with respect to the horizontal. What is the minimum initial speed that would allow the athlete to clear the gap? Neglect air resistance.

SOLUTION

1. DEFINE **Given:** $\Delta x = 6.0$ m

$\theta = 35°$

$a_y = -g = -9.81$ m/s^2

Unknown: $v_i = ?$

2. PLAN **Diagram:**

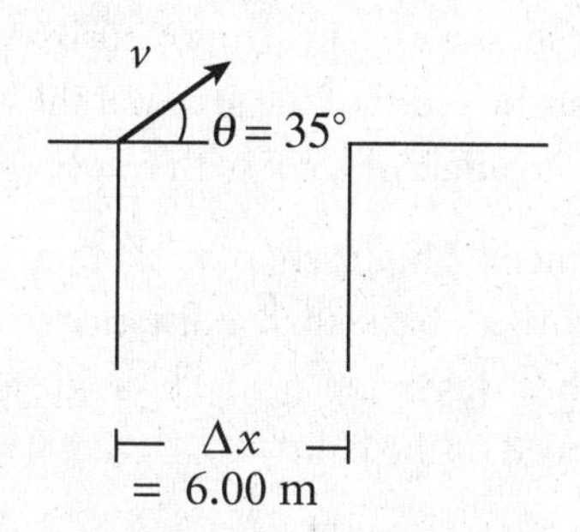

Choose the equation(s) or situation: The horizontal component of the athlete's velocity, v_x, is equal to the initial speed multiplied by the cosine of the angle, θ, which is equal to the magnitude of the horizontal displacement, Δx, divided by the time interval required for the complete jump.

$$v_x = v_i \cos \theta = \frac{\Delta x}{\Delta t}$$

At the midpoint of the jump, the vertical component of the athlete's velocity, v_y, which is the upward vertical component of the initial velocity, $v_i \sin \theta$, plus the component of velocity due to free-fall acceleration, equals zero. The time required for this to occur is half the time necessary for the total jump.

$$v_y = v_i \sin \theta + a_y\left(\frac{\Delta t}{2}\right) = 0$$

$$v_i \sin \theta = \frac{-a_y \Delta t}{2}$$

Rearrange the equation(s) to isolate the unknown(s): Express Δt in the second equation in terms of the displacement and velocity component in the first equation.

$$v_i \sin \theta = \frac{-a_y}{2}\left(\frac{\Delta x}{v_i \cos \theta}\right)$$

$$v_i^2 = \frac{-a_y \Delta x}{2 \sin \theta \cos \theta}$$

$$v_i = \sqrt{\frac{-a_y \Delta x}{2 \sin \theta \cos \theta}}$$

3. CALCULATE **Substitute the values into the equation(s) and solve:** Select the positive root for v_i.

$$v_i = \sqrt{\frac{-(-9.81 \text{ m/s}^2)(6.0 \text{ m})}{(2)(\sin 35°)(\cos 35°)}} = \boxed{7.9 \text{ m/s}}$$

4. EVALUATE By substituting the value for v_i into the original equations, you can determine the time for the jump to be completed, which is 0.92 s. From this, the height of the jump is found to equal 1.0 m.

ADDITIONAL PRACTICE

1. In 1993, Wayne Brian threw a spear a record distance of 201.24 m. (This is not an official sport record because a special device was used to "elongate" Brian's hand.) Suppose Brian threw the spear at a 35.0° angle with respect to the horizontal. What was the initial speed of the spear?
2. April Moon set a record in flight shooting (a variety of long-distance archery). In 1981 in Utah, she sent an arrow a horizontal distance of 9.50×10^2 m. What was the speed of the arrow at the top of the flight if the arrow was launched at an angle of 45.0° with respect to the horizontal?
3. In 1989 during overtime in a high school basketball game in Erie, Pennsylvania, Chris Eddy threw a basketball a distance of 27.5 m to score and win the game. If the shot was made at a 50.0° angle above the horizontal, what was the initial speed of the ball?
4. In 1978, Geoff Capes of the United Kingdom won a competition for throwing 5 lb bricks; he threw one brick a distance of 44.0 m. Suppose the brick left Capes' hand at an angle of 45.0° with respect to the horizontal.
   a. What was the initial speed of the brick?
   b. What was the maximum height reached by the brick?
   c. If Capes threw the brick straight up with the speed found in (a), what would be the maximum height the brick could achieve?
5. In 1991, Doug Danger rode a motorcycle to jump a horizontal distance of 76.5 m. Find the maximum height of the jump if his angle with respect to the ground at the beginning of the jump was 12.0°.
6. Michael Hout of Ohio can run 110.0 meter hurdles in 18.9 s at an average speed of 5.82 m/s. What makes this interesting is that he juggles three balls as he runs the distance. Suppose Hout throws a ball up and forward at twice his running speed and just catches it at the same level. At what angle, θ, must the ball be thrown? (Hint: Consider horizontal displacements for Hout and the ball.)

7. A scared kangaroo once cleared a fence by jumping with a speed of 8.42 m/s at an angle of 55.2° with respect to the ground. If the jump lasted 1.40 s, how high was the fence? What was the kangaroo's horizontal displacement?

8. Measurements made in 1910 indicate that the common flea is an impressive jumper, given its size. Assume that a flea's initial speed is 2.2 m/s, and that it leaps at an angle of 21° with respect to the horizontal. If the jump lasts 0.16 s, what is the magnitude of the flea's horizontal displacement? How high does the flea jump?

Two-Dimensional Motion and Vectors

Problem F

RELATIVE VELOCITY

PROBLEM

The world's fastest current is in Slingsby Channel, Canada, where the speed of the water reaches 30.0 km/h. Suppose a motorboat crosses the channel perpendicular to the bank at a speed of 18.0 km/h relative to the bank. Find the velocity of the motorboat relative to the water.

SOLUTION

1. DEFINE

Given: $\mathbf{v_{wb}}$ = 30.0 km/h along the channel (velocity of the *water, w,* with respect to the *bank, b)*

$\mathbf{v_{mb}}$ = 18.0 km/h perpendicular to the channel (velocity of the *motorboat, m,* with respect to the *bank, b)*

Unknown: $\mathbf{v_{mw}}$ = ?

Diagram:

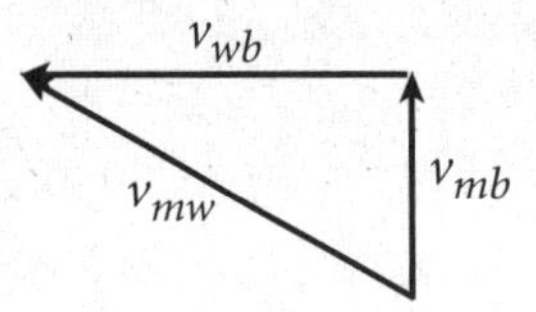

2. PLAN

Choose the equation(s) or situation: From the vector diagram, the resultant vector (the velocity of the motorboat with respect to the bank, $\mathbf{v_{mb}}$) is equal to the vector sum of the other two vectors, one of which is the unknown.

$$\mathbf{v_{mw}} = \mathbf{v_{mb}} + \mathbf{v_{wb}}$$

Use the Pythagorean theorem to calculate the magnitude of the resultant velocity, and use the tangent function to find the direction. Note that because the vectors $\mathbf{v_{mb}}$ and $\mathbf{v_{wb}}$ are perpendicular to each other, the product that results from multiplying one by the other is zero. The tangent of the angle between $\mathbf{v_{mb}}$ and $\mathbf{v_{mw}}$ is equal to the ratio of the magnitude of $\mathbf{v_{wb}}$ to the magnitude of $\mathbf{v_{mb}}$.

$$v_{mw}^2 = v_{mb}^2 + v_{wb}^2$$

$$\tan\theta = \frac{v_{wb}}{v_{mb}}$$

Rearrange the equation(s) to isolate the unknown(s):

$$v_{mw} = \sqrt{v_{mb}^2 + v_{wb}^2}$$

$$\theta = \tan^{-1}\left(\frac{v_{wb}}{v_{mb}}\right)$$

3. CALCULATE

Substitute the values into the equation(s) and solve: Choose the positive root for v_{mw}.

$$v_{mw} = \sqrt{\left(18.0\,\frac{\text{km}}{\text{h}}\right)^2 + \left(30.0\,\frac{\text{km}}{\text{h}}\right)^2} = \boxed{35.0\,\frac{\text{km}}{\text{h}}}$$

The angle between v_{mb} and v_{mw} is as follows:

$$\theta = \tan^{-1}\left(\frac{30.0\,\frac{\text{km}}{\text{h}}}{18.0\,\frac{\text{km}}{\text{h}}}\right) = \boxed{\text{59.0° away from the oncoming current}}$$

4. EVALUATE

The motorboat must move in a direction 59° with respect to $\mathbf{v_{mb}}$ and against the current, and with a speed of 35.0 km/h in order to move 18.0 km/h perpendicular to the bank.

ADDITIONAL PRACTICE

1. In 1933, a storm occurring in the Pacific Ocean moved with speeds reaching a maximum of 126 km/h. Suppose a storm is moving north at this speed. If a gull flies east through the storm with a speed of 40.0 km/h relative to the air, what is the velocity of the gull relative to Earth?
2. George V Coast in Antarctica is the windiest place on Earth. Wind speeds there can reach 3.00×10^2 km/h. If a research plane flies against the wind with a speed of 4.50×10^2 km/h relative to the wind, how long does it take the plane to fly between two research stations that are 250 km apart?
3. Turtles are fairly slow on the ground, but they are very good swimmers, as indicated by the reported speed of 9.0 m/s for the leatherback turtle. Suppose a leatherback turtle swims across a river at 9.0 m/s relative to the water. If the current in the river is 3.0 m/s and it moves at a right angle to the turtle's motion, what is the turtle's displacement with respect to the river's bank after 1.0 min?
4. California sea lions can swim as fast as 40.0 km/h. Suppose a sea lion begins to chase a fish at this speed when the fish is 60.0 m away. The fish, of course, does not wait, and swims away at a speed 16.0 km/h. How long would it take the sea lion to catch the fish?
5. The spur-wing goose is one of the fastest birds in the world when it comes to *level* flying: it can reach a speed of 90.0 km/h. Suppose two spur-wing geese are separated by an unknown distance and start flying toward each other at their maximum speeds. The geese pass each other 40.0 s later. Calculate the initial distance between the geese.
6. The fastest snake on Earth is the black mamba, which can move over a short distance at 18.0 km/h. Suppose a mamba moves at this speed toward a rat sitting 12.0 m away. The rat immediately begins to run away at 33.3 percent of the mamba's speed. If the rat jumps into a hole just before the mamba can catch it, determine the length of time that the chase lasts.

Forces and the Laws of Motion

Problem A

DRAWING FREE-BODY DIAGRAMS

PROBLEM

In the early morning, a park ranger in a canoe is observing wildlife on the nearby shore. The Earth's gravitational force on the ranger is 760 N downward and its gravitational force on the boat is 190 N downward. The water keeps the canoe afloat by exerting a 950 N force upward on it. Draw a free-body diagram of the canoe.

SOLUTION

1. **Identify the forces acting on the object and the directions of the forces.**
   - The Earth exerts a force of 190 N downward on the canoe.
   - The park ranger exerts a force of 760 N downward on the canoe.
   - The water exerts an upward force of 950 N on the canoe.
2. **Draw a diagram to represent the isolated object.**
   The canoe can be represented by a simple outline, as shown in **(a)**.
3. **Draw and label vector arrows for all external forces acting on the object.**
   A free-body diagram of the canoe will show all the forces acting on the canoe as if the forces are acting on the center of the canoe. First, draw and label the gravitational force acting on the canoe, which is directed toward the center of Earth, as shown in **(b)**. Be sure that the length of the arrow approximately represents the magnitude of the force.

   Next, draw and label the downward force that is exerted on the boat by the Earth's gravitational attraction on the ranger, as shown in **(c)**. Finally, draw and label the upward force exerted by the water on the canoe as shown in **(d)**. Diagram **(d)** is the completed free-body diagram of the floating canoe.

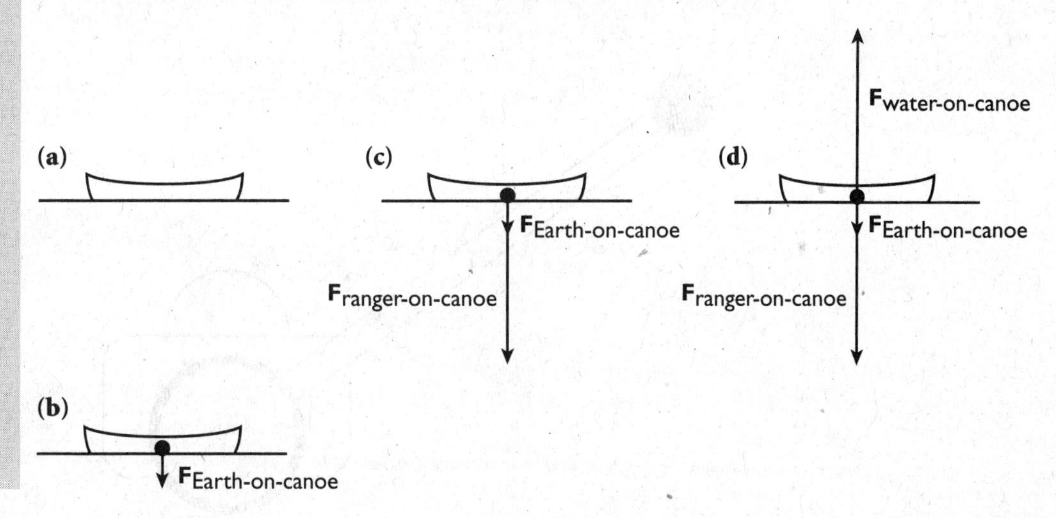

ADDITIONAL PRACTICE

1. After a skydiver jumps from a plane, the only force initially acting on the diver is Earth's gravitational attraction. After about ten seconds of falling, air resistance on the diver will have increased so that its magnitude on the diver is now equal in magnitude to Earth's gravitational force on the diver. At this time, a diver in a belly-down position will be falling at a constant speed of about 190 km/h.

a. Draw a free-body diagram of the skydiver when the diver initially leaves the plane.

b. Draw a free-body diagram of the skydiver at the tenth second of the falling.

2. A chef places an open sack of flour on a kitchen scale. The scale reading of 40 N indicates that the scale is exerting an upward force of 40 N on the sack. The magnitude of this force equals the magnitude of the force of Earth's gravitational attraction on the sack. The chef then exerts an upward force of 10 N on the bag and the scale reading falls to 30 N. Draw a free-body diagram of the latter situation.

3. A music box within the toy shown below plays tunes when the toy is pushed along the floor. As a child pushes along the handlebars with a force of 5 N, the floor exerts a force of 13 N directly upward on the toy. The Earth's gravitational force on the toy is 10 N downward while interactions between the wheels and the floor produce a backward force of 2 N on the toy as it moves. Draw a free-body diagram of the toy as it is being pushed.

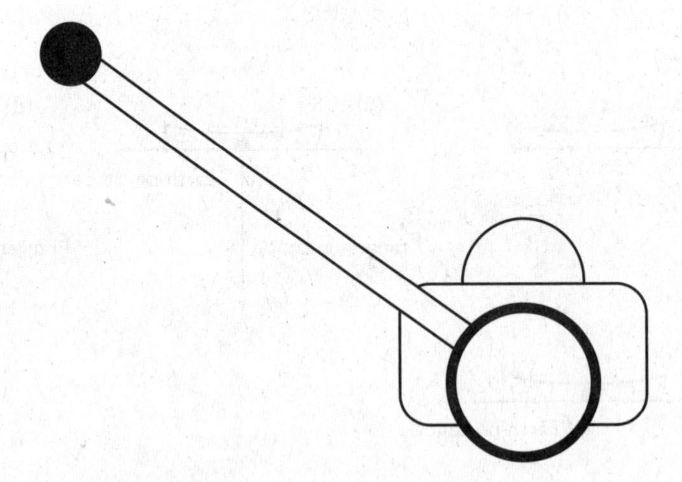

Forces and the Laws of Motion

Problem B

DETERMINING NET FORCE

PROBLEM

The muscle responsible for closing the mouth is the strongest muscle in the human body. It can exert a force greater than that exerted by a man lifting a mass of 400 kg. Richard Hoffman of Florida recorded the force of biting at 4.33×10^3 N. If each force shown in the diagram below has a magnitude equal to the force of Hoffman's bite, determine the net force.

SOLUTION

1. DEFINE **Define the problem, and identify the variables.**

Given: $F_1 = -4.33 \times 10^3$ N

$F_2 = 4.33 \times 10^3$ N

$F_3 = 4.33 \times 10^3$ N

Unknown: $F_{net} = ?$ $\theta_{net} = ?$

Diagram:

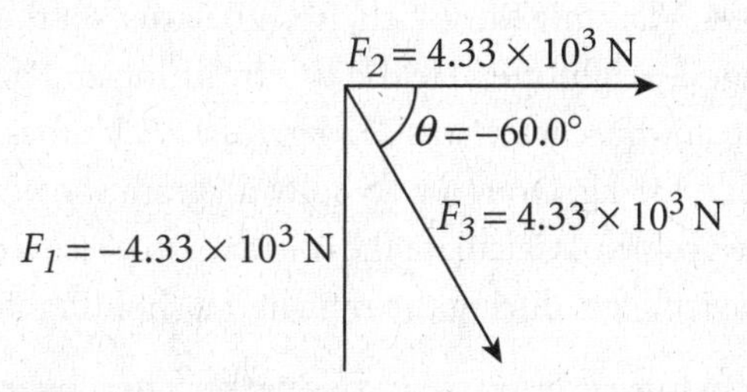

2. PLAN **Select a coordinate system, and apply it to the free-body diagram:** Let F_1 lie along the negative y-axis and F_2 lie along the positive x-axis. Now F_3 must be resolved into x and y components.

3. CALCULATE **Find the x and y components of all vectors:** As indicated in the sketch, the angle between F_3 and the x-axis is 60.0°. Because this angle is in the quadrant bounded by the positive x and negative y axes, it has a negative value.

$$F_{3,x} = F_3(\cos\theta) = (4.33 \times 10^3 \text{ N})\,[\cos(-60.0°)] = 2.16 \times 10^3 \text{ N}$$
$$F_{3,y} = F_3(\sin\theta) = (4.33 \times 10^3 \text{ N})\,[\sin(-60.0°)] = -3.75 \times 10^3 \text{ N}$$

Find the net external force in both the x and y directions.

For the x direction: $\Sigma F_x = F_2 + F_{3,x} = F_{x,net}$

$$\Sigma F_x = 4.33 \times 10^3 \text{ N} + 2.16 \times 10^3 \text{ N} = 6.49 \times 10^3 \text{ N}$$

For the y direction: $\Sigma F_y = F_1 + F_{3,y} = F_{y,net}$

$$\Sigma F_y = (-4.33 \times 10^3 \text{ N}) + (-3.75 \times 10^3 \text{ N}) = -8.08 \times 10^3 \text{ N}$$

Find the net external force.

Use the Pythagorean theorem to calculate F_{net}. Use $\theta_{net} = \tan^{-1}\left(\frac{F_{y,net}}{F_{x,net}}\right)$ to find the angle between the net force and the x-axis.

$$F_{net} = \sqrt{(F_{x,net})^2 + (F_{y,net})^2}$$
$$F_{net} = \sqrt{(6.49 \times 10^3 \text{ N})^2 + (-8.08 \times 10^3 \text{ N})^2} = \sqrt{10.74 \times 10^7 \text{ N}^2}$$
$$F_{net} = \boxed{1.036 \times 10^4 \text{ N}}$$

$$\theta_{net} = \tan^{-1}\left(\frac{-8.08 \times 10^3\ \text{N}}{6.49 \times 10^3\ \text{N}}\right) = \boxed{-51.2°}$$

4. EVALUATE The net force is larger than the individual forces, but it is not quite three times as large as any one force, which would be the case if all three forces were acting in one direction only. The angle is negative to indicate that it is in the quadrant below the positive x-axis, where the values along the y-axis are negative. The net force is 1.036×10^4 N at an angle of 51.2° below the positive x-axis.

ADDITIONAL PRACTICE

1. Joe Ponder, from North Carolina, once used his teeth to lift a pumpkin with a mass of 275 kg. Suppose Ponder has a mass of 75 kg, and he stands with each foot on a platform and lifts the pumpkin with an attached rope. If he holds the pumpkin above the ground between the platforms, what is the force exerted on his feet? (Draw a free-body diagram showing all of the forces present on Ponder.)

2. In 1994, Vladimir Kurlovich, from Belarus, set the record as the world's strongest weightlifter. He did this by lifting and holding above his head a barbell whose mass was 253 kg. Kurlovich's mass at the time was roughly 133 kg. Draw a free-body diagram showing the various forces in the problem. Calculate the normal force exerted on *each* of Kurlovich's feet during the time he was holding the barbell.

3. The net force exerted by a woodpecker's head when its beak strikes a tree can be as large as 4.90 N, assuming that the bird's head has a mass of 50.0 g. Assume that two different muscles pull the woodpecker's head forward and downward, exerting a net force of 4.90 N. If the forces exerted by the muscles are at right angles to each other and the muscle that pulls the woodpecker's head downward exerts a force of 1.70 N, what is the magnitude of the force exerted by the other muscle? Draw a free-body diagram showing the forces acting on the woodpecker's head.

4. About 50 years ago, the San Diego Zoo, in California, had the largest gorilla on Earth: its mass was about 3.10×10^2 kg. Suppose a gorilla with this mass hangs from two vines, each of which makes an angle of 30.0° with the vertical. Draw a free-body diagram showing the various forces, and find the magnitude of the force of tension in each vine. What would happen to the tensions if the upper ends of the vines were farther apart?

5. The mass of Zorba, a mastiff born in London, England, was measured in 1989 to be 155 kg. This mass is roughly the equivalent of the combined masses of two average adult male mastiffs. Suppose Zorba is placed in a harness that is suspended from the ceiling by two cables that are at right angles to each other. If the tension in one cable is twice as large as the tension in the other cable, what are the magnitudes of the two tensions? Assume the mass of the cables and harness to be negligible. Before doing the calculations, draw a free-body diagram showing the forces acting on Zorba.

NAME ______________________ DATE ______________ CLASS ______________

Forces and the Laws of Motion

Problem C

NEWTON'S SECOND LAW

PROBLEM

A 1.5 kg ball has an acceleration of 9.0 m/s² to the left. What is the net force acting on the ball?

SOLUTION

Given: $m = 1.5$ kg

$\mathbf{a} = 9.0 \text{ m/s}^2$ to the left

Unknown: $\mathbf{F} = ?$

Use Newton's second law, and solve for **F**.

$$\Sigma \mathbf{F} = m\mathbf{a}$$

Because there is only one force,

$$\Sigma \mathbf{F} = \mathbf{F}$$

$$F = (1.5 \text{ kg})(9.0 \text{ m/s}^2) = 14 \text{ N}$$

$$\mathbf{F} = \boxed{14 \text{ N to the left}}$$

ADDITIONAL PRACTICE

1. David Purley, a racing driver, survived deceleration from 173 km/h to 0 km/h over a distance of 0.660 m when his car crashed. Assume that Purley's mass is 70.0 kg. What is the average force acting on him during the crash? Compare this force to Purley's weight. (Hint: Calculate the average acceleration first.)

2. A giant crane in Washington, D. C. was tested by lifting a 2.232×10^6 kg load.

a. Find the magnitude of the force needed to lift the load with a net acceleration of 0 m/s².

b. If the same force is applied to pull the load up a smooth slope that makes a 30.0° angle with the horizontal, what would be the acceleration?

3. When the click beetle jumps in the air, its acceleration upward can be as large as 400.0 times the acceleration due to gravity. (An acceleration this large would instantly kill any human being.) For a beetle whose mass is 40.00 mg, calculate the magnitude of the force exerted by the beetle on the ground at the beginning of the jump with gravity taken into account. Calculate the magnitude of the force with gravity neglected. Use 9.807 m/s² as the value for free-fall acceleration.

4. In 1994, a Bulgarian athlete named Minchev lifted a mass of 157.5 kg. By comparison, his own mass was only 54.0 kg. Calculate the force acting on each of his feet at the moment he was lifting the mass with an upward acceleration of 1.00 m/s^2. Assume that the downward force on each foot is the same.

5. In 1967, one of the high school football teams in California had a tackle named Bob whose mass was 2.20×10^2 kg. Suppose that after winning a game the happy teammates throw Bob up in the air but fail to catch him. When Bob hits the ground, his average upward acceleration over the course of the collision is 75.0 m/s^2. (Note that this acceleration has a much greater magnitude than free-fall acceleration.) Find the average force that the ground exerts on Bob during the collision.

6. The whale shark is the largest type of fish in the world. Its mass can be as large as 2.00×10^4 kg, which is the equivalent mass of three average adult elephants. Suppose a crane lifts a net with a 2.00×10^4 kg whale shark off the ground. The net is steadily accelerated from rest over an interval of 2.5 s until the net reaches a speed of 1.0 m/s. Calculate the magnitude of the tension in the cable pulling the net.

7. The largest toad ever caught had a mass of 2.65 kg. Suppose a toad with this mass is placed on a metal plate that is attached to two cables, as shown in the figure below. If the plate is pulled upward so that it has a net acceleration of 2.55 m/s^2, what is magnitude of the tension in the cables? (The plate's weight can be disregarded.)

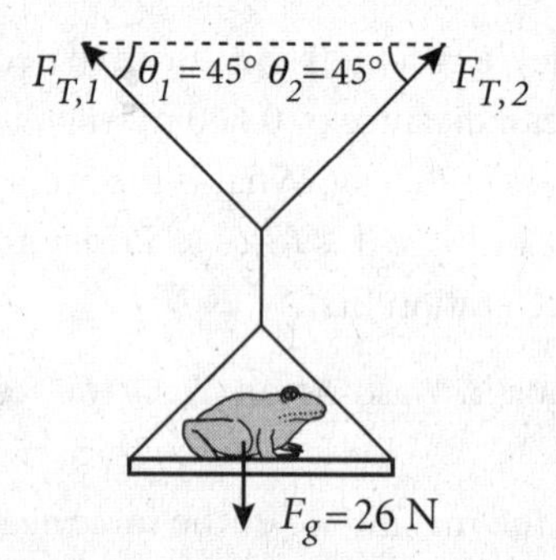

8. In 1991, a lobster with a mass of 20.0 kg was caught off the coast of Nova Scotia, Canada. Imagine this lobster involved in a friendly tug of war with several smaller lobsters on a horizontal plane at the bottom of the sea. Suppose the smaller lobsters are able to drag the large lobster, so that after the large lobster has been moved 1.55 m its speed is 0.550 m/s. If the lobster is initially at rest, what is the magnitude of the net force applied to it by the smaller lobsters? Assume that friction and resistance due to moving through water are negligible.

9. A 0.5 mm wire made of carbon and manganese can just barely support the weight of a 70.0 kg person. Suppose this wire is used to lift a 45.0 kg load. What maximum upward acceleration can be achieved without breaking the wire?

10. The largest hydraulic turbines in the world have shafts with individual masses that equal 3.18×10^5 kg. Suppose such a shaft is delivered to the assembly line on a trailer that is pulled with a horizontal force of 81.0 kN. If the force of friction opposing the motion is 62.0 kN, what is the magnitude of the trailer's net acceleration? (Disregard the mass of the trailer.)

11. An average newborn blue whale has a mass of 3.00×10^3 kg. Suppose the whale becomes stranded on the shore and a team of rescuers tries to pull it back to sea. The rescuers attach a cable to the whale and pull it at an angle of 20.0° above the horizontal with a force of 4.00 kN. There is, however, a horizontal force opposing the motion that is 12.0 percent of the whale's weight. Calculate the magnitude of the whale's net acceleration during the rescue pull.

12. One end of the cable of an elevator is attached to the elevator car, and the other end of the cable is attached to a counterweight. The counterweight consists of heavy metal blocks with a total mass almost the same as the car's. By using the counterweight, the motor used to lift and lower the car needs to exert a force that is only about equal to the total weight of the passengers in the car. Suppose the car with passengers has a mass of 1.600×10^3 kg and the counterweight has a mass of 1.200×10^3 kg. Calculate the magnitude of the car's net acceleration as it falls from rest at the top of the shaft to the ground 25.0 m below. Calculate the car's final speed.

13. The largest squash ever grown had a mass of 409 kg. Suppose you want to push a squash with this mass up a smooth ramp that is 6.00 m long and that makes a 30.0° angle with the horizontal. If you push the squash with a force of 2080 N up the incline, what is

a. the net force exerted on the squash?

b. the net acceleration of the squash?

c. the time required for the squash to reach the top of the ramp?

14. A very thin boron rod with a cross-section of 0.10 mm × 0.10 mm can sustain a force of 57 N. Assume the rod is used to pull a block along a smooth horizontal surface.

a. If the maximum force accelerates the block by 0.25 m/s^2, find the mass of the block.

b. If a second force of 24 N is applied in the direction opposite the 57 N force, what would be the magnitude of the block's new acceleration?

15. A hot-air balloon with a total mass of 2.55×10^3 kg is being pulled down by a crew tugging on a rope. The tension in the rope is 7.56×10^3 N at an angle of 72.3° below the horizontal. This force is aided in the vertical direction by the balloon's weight and is opposed by a buoyant force of 3.10×10^4 N that lifts the balloon upward. A wind blowing from behind the crew exerts a horizontal force of 920 N on the balloon.

a. What is the magnitude and direction of the net force?

b. Calculate the magnitude of the balloon's net acceleration.

c. Suppose the balloon is 45.0 m above the ground when the crew begins pulling it down. How far will the balloon travel horizontally by the time it reaches the ground if the balloon is initially at rest?

NAME ______________________ DATE ______________ CLASS ______________

Forces and the Laws of Motion

Problem D

COEFFICIENTS OF FRICTION

PROBLEM

A 20.0 kg trunk is pushed across the floor of a moving van by a horizontal force. If the coefficient of kinetic friction between the trunk and the floor is 0.255, what is the magnitude of the frictional force opposing the applied force?

SOLUTION

Given: $m = 20.0$ kg

$\mu_k = 0.255$

$g = 9.81 \text{ m/s}^2$

Unknown: $F_k = ?$

Use the equation for frictional force, substituting mg for the normal force F_n.

$$F_k = \mu_k F_n = \mu_k mg$$

$$F_k = (0.255)(20.0 \text{ kg})(9.81 \text{ m/s}^2)$$

$$F_k = \boxed{50.0 \text{ N}}$$

ADDITIONAL PRACTICE

1. The largest flowers in the world are the *Rafflesia arnoldii*, found in Malaysia. A single flower is almost a meter across and has a mass up to 11.0 kg. Suppose you cut off a single flower and drag it along the flat ground. If the coefficient of kinetic friction between the flower and the ground is 0.39, what is the magnitude of the frictional force that must be overcome?

2. Robert Galstyan, from Armenia, pulled two coupled railway wagons a distance of 7 m using his teeth. The total mass of the wagons was about 2.20×10^5 kg. Of course, his job was made easier by the fact that the wheels were free to roll. Suppose the wheels are blocked and the coefficient of static friction between the rails and the sliding wheels is 0.220. What would be the magnitude of the minimum force needed to move the wagons from rest? Assume that the track is horizontal.

3. The steepest street in the world is Baldwin Street in Dunedin, New Zealand. It has an inclination angle of 38.0° with respect to the horizontal. Suppose a wooden crate with a mass of 25.0 kg is placed on Baldwin Street. An additional force of 59 N must be applied to the crate perpendicular to the pavement in order to hold the crate in place. If the coefficient of static friction between the crate and the pavement is 0.599, what is the magnitude of the frictional force?

4. Now imagine that a child rides a wagon down Baldwin Street. In order to keep from moving too fast, the child has secured the wheels of the wagon so that they do not turn. The wagon and child then slide down the hill at a constant velocity. What is the coefficient of kinetic friction between the tires of the wagon and the pavement?

5. The steepest railroad track that allows trains to move using their own locomotion and the friction between their wheels and the track is located in France. The track makes an angle of 5.2° with the horizontal. Suppose the rails become greasy and the train slides down the track even though the wheels are locked and held in place with blocks. If the train slides down the tracks with a constant velocity, what is the coefficient of kinetic friction between the wheels and track?

6. Walter Arfeuille of Belgium lifted a 281.5 kg load off the ground using his teeth. Suppose Arfeuille can hold just three times that mass on a 30.0° slope using the same force. What is the coefficient of static friction between the load and the slope?

7. A blue whale with a mass of 1.90×10^5 kg was caught in 1947. What is the magnitude of the minimum force needed to move the whale along a horizontal ramp if the coefficient of static friction between the ramp's surface and the whale is 0.460?

8. Until 1979, the world's easiest driving test was administered in Egypt. To pass the test, one needed only to drive about 6 m forward, stop, and drive the same distance in reverse. Suppose that at the end of the 6 m the car's brakes are suddenly applied and the car slides to a stop. If the force required to stop the car is 6.0×10^3 N and the coefficient of kinetic friction between the tires and pavement is 0.77, what is the magnitude of the car's normal force? What is the car's mass?

9. The heaviest train ever pulled by a single engine was over 2 km long. Suppose a force of 1.13×10^8 N is needed to overcome static friction in the train's wheels. If the coefficient of static friction is 0.741, what is the train's mass?

10. In 1994, a 3.00×10^3 kg pancake was cooked and flipped in Manchester, England. If the pancake is placed on a surface that is inclined 31.0° with respect to the horizontal, what must the coefficient of kinetic friction be in order for the pancake to slide down the surface with a constant velocity? What would be the magnitude of the frictional force acting on the pancake?

Forces and the Laws of Motion

Problem E

OVERCOMING FRICTION

PROBLEM

In 1988, a very large telephone constructed by a Dutch telecommunications company was demonstrated in the Netherlands. Suppose this telephone is towed a short distance by a horizontal force equal to 8670 N, so that the telephone's net acceleration is 1.30 m/s^2. Given that the coefficient of kinetic friction between the phone and the ground is 0.120, calculate the mass of the telephone.

SOLUTION

1. DEFINE

Given: $F_{applied} = 8670$ N

$a_{net} = 1.30$ m/s^2

$\mu_k = 0.120$

$g = 9.81$ m/s^2

Unknown: $m = ?$

2. PLAN

Choose the equation(s) or situation: Apply Newton's second law to describe the forces acting on the telephone.

$$F_{net} = ma_{net} = F_{applied} - F_k$$

The frictional force, F_k, depends on the normal force, F_n, exerted on the telephone. For a horizontal surface, the normal force equals the telephone's weight.

$$F_k = \mu_k F_n = \mu_k(mg)$$

Substituting the equation for F_k into the equation for F_{net} provides an equation with all known and unknown variables.

$$ma_{net} = F_{applied} - \mu_k mg$$

Rearrange the equation(s) to isolate the unknown(s):

$$ma_{net} + \mu_k mg = F_{applied}$$

$$m(a_{net} + \mu_k g) = F_{applied}$$

$$m = \frac{F_{applied}}{a_{net} + \mu_k g}$$

3. CALCULATE

Substitute the values into the equations and solve:

$$m = \frac{8670 \text{ N}}{1.30 \text{ m/s}^2 + (0.120)(9.81 \text{ m/s}^2)}$$

$$m = \frac{8670 \text{ N}}{1.30 \text{ m/s}^2 + 1.18 \text{ m/s}^2}$$

$$m = \frac{8670 \text{ N}}{2.48 \text{ m/s}^2} = \boxed{3.50 \times 10^3 \text{ kg}}$$

4. EVALUATE

Because the mass is constant, the sum of the acceleration terms in the denominator must be constant for a constant applied force. Therefore, the net acceleration decreases as the coefficient of friction increases.

ADDITIONAL PRACTICE

1. Isaac Newton developed the laws of mechanics. Brian Newton put those laws into application when he ran a marathon in about 8.5 h while carrying a bag of coal. Suppose Brian Newton wants to remove the bag from the finish line. He drags the bag with an applied horizontal force of 130 N, so that the bag as a net acceleration of 1.00 m/s^2. If the coefficient of kinetic friction between the bag and the pavement is 0.158, what is the mass of the bag of coal?

2. The most massive car ever built was the official car of the General Secretary of the Communist Party in the former Soviet Union. Suppose this car is moving down a 10.0° slope when the driver suddenly applies the brakes. The net force acting on the car as it stops is -2.00×10^4 N. If the coefficient of kinetic friction between the car's tires and the pavement is 0.797, what is the car's mass? What is the magnitude of the normal force that the pavement exerts on the car?

3. Suppose a giant hamburger slides down a ramp that has a 45.0° incline. The coefficient of kinetic friction between the hamburger and the ramp is 0.597, so that the net force acting on the hamburger is 6.99×10^3 N. What is the mass of the hamburger? What is the magnitude of the normal force that the ramp exerts on the hamburger?

4. An extremely light, drivable car with a mass of only 9.50 kg was built in London. Suppose that the wheels of the car are locked, so that the car no longer rolls. If the car is pushed up a 30.0° slope by an applied force of 80.0 N, the net acceleration of the car is 1.64 m/s^2. What is the coefficient of kinetic friction between the car and the incline?

5. *Cleopatra's Needle*, an obelisk given by the Egyptian government to Great Britain in the nineteenth century, is 20+ m tall and has a mass of about 1.89×10^5 kg. Suppose the monument is lowered onto its side and dragged horizontally to a new location. An applied force of 7.6×10^5 N is exerted on the monument, so that its net acceleration is 0.11 m/s^2. What is the magnitude of the frictional force?

6. Snowfall is extremely rare in Dunedin, New Zealand. Nevertheless, suppose that Baldwin Street, which has an incline of 38.0°, is covered with snow and that children are sledding down the street. A sled and rider move downhill with a constant acceleration. What would be the magnitude of the sled's net acceleration if the coefficient of kinetic friction between the snow and the sled's runners is 0.100? Does the acceleration depend on the masses of the sled and rider?

7. The record speed for grass skiing was set in 1985 by Klaus Spinka, of Austria. Suppose it took Spinka 6.60 s to reach his top speed after he started from rest down a slope with a 34.0° incline. If the coefficient of kinetic friction between the skis and the grass was 0.198, what was the magnitude of Spinka's net acceleration? What was his speed after 6.60 s?

Work and Energy

Problem A

WORK

PROBLEM

The largest palace in the world is the Imperial Palace in Beijing, China. Suppose you were to push a lawn mower around the perimeter of a rectangular area identical to that of the palace, applying a constant horizontal force of 60.0 N. If you did 2.05×10^5 J of work, how far would you have pushed the lawn mower? If the Imperial Palace is 9.60×10^2 m long, how wide is it?

SOLUTION

Given: $F = 60.0$ N

$W = 2.05 \times 10^5$ J

$x = 9.60 \times 10^2$ m

Unknown: $d = ?$

$y = ?$

Use the equation for net work done by a constant force.

$$W = Fd(\cos\theta)$$

To calculate the width, y, recall that the perimeter of an area equals the sum of twice its width and twice its length.

$$d = 2x + 2y$$

Rearrange the equations to solve for d and y. Note that the force is applied in the direction of the displacement, so $\theta = 0°$.

$$d = \frac{W}{F(\cos\theta)} = \frac{2.05 \times 10^5 \text{ J}}{(60.0 \text{ N})(\cos 0°)}$$

$$d = \boxed{3.42 \times 10^3 \text{ m}}$$

$$y = \frac{d - 2x}{2} = \frac{3.42 \times 10^3 \text{ m} - (2)(9.60 \times 10^2 \text{ m})}{2}$$

$$y = \frac{3.42 \times 10^3 \text{m} - 1.92 \times 10^3 \text{ m}}{2} = \frac{1.50 \times 10^3 \text{ m}}{2}$$

$$y = \boxed{7.50 \times 10^2 \text{ m}}$$

ADDITIONAL PRACTICE

1. Lake Point Tower in Chicago is the tallest apartment building in the United States (although not the tallest building in which there are apartments). Suppose you take the elevator from street level to the roof of the building. The elevator moves almost the entire distance at constant speed, so that it does 1.15×10^5 J of work on you as it lifts the

entire distance. If your mass is 60.0 kg, how tall is the building? Ignore the effects of friction.

2. In 1985 in San Antonio, Texas, an entire hotel building was moved several blocks on 36 dollies. The mass of the building was about 1.45×10^6 kg. If 1.00×10^2 MJ of work was done to overcome the force of resistance that was just 2.00 percent of the building's weight, how far was the building moved?

3. A hummingbird has a mass of about 1.7 g. Suppose a hummingbird does 0.15 J of work against gravity, so that it ascends straight up with a net acceleration of 1.2 m/s^2. How far up does it move?

4. In 1453, during the siege of Constantinople, the Turks used a cannon capable of launching a stone cannonball with a mass of 5.40×10^2 kg. Suppose a soldier dropped a cannonball with this mass while trying to load it into the cannon. The cannonball rolled down a hill that made an angle of 30.0° with the horizontal. If 5.30×10^4 J of work was done by gravity on the cannonball as it rolled down a hill, how far did it roll?

5. The largest turtle ever caught in the United States had a mass of over 800 kg. Suppose this turtle were raised 5.45 m onto the deck of a research ship. If it takes 4.60×10^4 J of work to lift the turtle this distance at a constant velocity, what is the turtle's weight?

6. During World War II, 16 huge wooden hangers were built for United States Navy airships. The hangars were over 300 m long and had a maximum height of 52.0 m. Imagine a 40.0 kg block being lifted by a winch from the ground to the top of the hangar's ceiling. If the winch does 2.08×10^4 J of work in lifting the block, what force is exerted on the block?

7. The *Warszawa Radio* mast in Warsaw, Poland, is 646 m tall, making it the tallest human-built structure. Suppose a worker raises some tools to the top of the tower by means of a small elevator. If 2.15×10^5 J of work is done in lifting the tools, what is the force exerted on them?

8. The largest mincemeat pie ever created had a mass of 1.02×10^3 kg. Suppose that a pie with this mass slides down a ramp that is 18.0 m long and is inclined to the ground by 10.0°. If the coefficient of kinetic friction is 0.13, what is the net work done on the pie during its descent?

9. The longest shish kebab ever made was 881.0 m long. Suppose the meat and vegetables need to be delivered in a cart from one end of this shish kebab's skewer to the other end. A cook pulls the cart by applying a force of 40.00 N at an angle of 45.00° above the horizontal. If the force of friction acting on the cart is 28.00 N, what is the net work done on the cart and its contents during the delivery?

10. The world's largest chandelier was created by a company in South Korea and hangs in one of the department stores in Seoul, South Korea. The chandelier's mass is about 9.7×10^3 kg. Consider a situation in which this chandelier is placed in a wooden crate whose mass is negligible. The chandelier is then pulled along a smooth horizontal surface by two forces that are parallel to the smooth surface, are at right angles to each other, and are applied 45° to either side of the direction in which the chandelier is moving. If each of these forces is 1.2×10^3 N, how much work must be done on the chandelier to pull it 12 m?

11. The world's largest flag, which was manufactured in Pennsylvania, has a length of 154 m and a width of 78 m. The flag's mass is 1.24×10^3 kg, which may explain why the flag has never been flown from a flagpole. Suppose this flag is being pulled by two forces: a force of 8.00×10^3 N to the east and a force of 5.00×10^3 N that is directed 30.0° south of east. How much work is done in moving the flag 20.0 m directly south?

Work and Energy

Problem B

KINETIC ENERGY

PROBLEM

Silvana Cruciata from Italy set a record in one-hour running by running 18.084 km in 1.000 h. If Cruciata's kinetic energy was 694 J, what was her mass?

SOLUTION

1. DEFINE **Given:**

$\Delta x = 18.084 \text{ km} = 1.8084 \times 10^4 \text{ m}$

$\Delta t = 1.000 \text{ h} = 3.600 \times 10^3 \text{ s}$

$KE = 694 \text{ J}$

Unknown: $m = ?$

2. PLAN **Choose the equation(s) or situation:** Use the definition of average velocity to calculate Cruciata's speed.

$$v_{avg} = \frac{\Delta x}{\Delta t}$$

Use the equation for kinetic energy, using v_{avg} for the velocity term, to solve for m.

$$KE = \frac{1}{2} m v_{avg}^2$$

Rearrange the equation(s) to isolate the unknown(s): Substitute the average velocity equation into the equation for kinetic energy and solve for m.

$$m = \frac{2KE}{v_{avg}^2} = \frac{2KE}{\left(\frac{\Delta x}{\Delta t}\right)^2} = \frac{2KE\Delta t^2}{\Delta x^2}$$

3. CALCULATE **Substitute the values into the equation(s) and solve:**

$$m = \frac{(2)(694 \text{ J})(3.600 \times 10^3 \text{ s})^2}{(1.8084 \times 10^4 \text{ m})^2} = \boxed{55.0 \text{ kg}}$$

4. EVALUATE If the average speed is rounded to 5.0 m/s, and the kinetic energy is rounded to 700 J, the estimated mass is 56 kg, which is close to the calculated value.

ADDITIONAL PRACTICE

1. In 1994, Leroy Burrell of the United States set what was then a new world record for the men's 100 m run. He ran the 1.00×10^2 m distance in 9.85 s. Assuming that he ran with a constant speed equal to his average speed, and his kinetic energy was 3.40×10^3 J, what was Burrell's mass?
2. The fastest helicopter, the Westland Lynx, has a top speed of 4.00×10^2 km/h. If its kinetic energy at this speed is 2.10×10^7 J, what is the helicopter's mass?

3. Dan Jansen of the United States won a speed-skating competition at the 1994 Winter Olympics in Lillehammer, Norway. He did this by skating 500 m with an average speed of 50.3 km/h. If his kinetic energy was 6.54×10^3 J, what was his mass?

4. In 1987, the fastest auto race in the United States was the Busch Clash in Daytona, Florida. That year, the winner's average speed was about 318 km/h. Suppose the kinetic energy of the winning car was 3.80 MJ. What was the mass of the car and its driver?

5. In 1995, Karine Dubouchet of France reached a record speed in downhill skiing. If Dubouchet's mass was 51.0 kg, her kinetic energy would have been 9.96×10^4 J. What was her speed?

6. Susie Maroney from Australia set a women's record in long-distance swimming by swimming 93.625 km in 24.00 h.

a. What was Maroney's average speed?

b. If Maroney's mass was 55 kg, what was her kinetic energy?

7. The brightest, hottest, and most massive stars are the brilliant blue stars designated as spectral class O. If a class O star with a mass of 3.38×10^{31} kg has a kinetic energy of 1.10×10^{42} J, what is its speed? Express your answer in km/s (a typical unit for describing the speed of stars).

8. The male polar bear is the largest land-going predator. Its height when standing on its hind legs is over 3 m and its mass, which is usually around 500 kg, can be as large as 680 kg. In spite of this bulk, a running polar bear can reach speeds of 56.0 km/h.

a. Determine the kinetic energy of a running polar bear, using the maximum values for its mass and speed.

b. What is the ratio of the polar bear's kinetic energy to the kinetic energy of Leroy Burrell, as given in item 1?

9. Escape speed is the speed required for an object to leave Earth's orbit. It is also the minimum speed an incoming object must have to avoid being captured and pulled into an orbit around Earth. The escape speed for a projectile launched from Earth's surface is 11.2 km/s. Suppose a meteor is pulled toward Earth's surface and, as a meteorite, strikes the ground with a speed equal to this escape speed. If the meteorite has a diameter of about 3 m and a mass of 2.3×10^5 kg, what is its kinetic energy at the instant it collides with Earth's surface?

NAME ______________________ DATE ______________ CLASS ______________

Work and Energy

Problem C

WORK-KINETIC ENERGY THEOREM

PROBLEM

The Great Pyramid of Khufu in Egypt, used to have a height of 147 m and sides that sloped at an angle of 52.0° with respect to the ground. Stone blocks with masses of 1.37×10^4 kg were used to construct the pyramid. Suppose that a block with this mass at rest on top of the pyramid begins to slide down the side. Calculate the block's kinetic energy at ground level if the coefficient of kinetic friction is 0.45.

SOLUTION

1. DEFINE **Given:**

$m = 1.37 \times 10^4$ kg

$h = 147$ m

$g = 9.81$ m/s^2

$\theta = 52.0°$

$\mu_k = 0.45$

$v_i = 0$ m/s

Unknown: $KE_f = ?$

2. PLAN **Choose the equation(s) or situation:** The net work done by the block as it slides down the side of the pyramid can be expressed by using the definition of work in terms of net force. Because the net force is parallel to the displacement, the net work is simply the net force multiplied by the displacement. It can also be expressed in terms of changing kinetic energy by using the work–kinetic energy theorem.

$$W_{net} = F_{net}d$$
$$W_{net} = \Delta KE$$

The net force on the block equals the difference between the component of the force due to free-fall acceleration along the side of the pyramid and the frictional force resisting the downward motion of the block.

$$F_{net} = mg(\sin\theta) - F_k = mg(\sin\theta) - \mu_k mg(\cos\theta)$$

The distance the block travels along the side of the pyramid equals the height of the pyramid divided by the sine of the angle of the side's slope.

$$h = d(\sin\theta)$$
$$d = \frac{h}{\sin\theta}$$

Because the block is initially at rest, its initial kinetic energy is zero, and the change in kinetic energy equals the final kinetic energy.

$$\Delta KE = KE_f - KE_i = KE_f$$

Rearrange the equation(s) to isolate the unknown(s): Combining these equations yields the following expression for the final kinetic energy.

$$KE_f = F_{net}d = mg(\sin\theta - \mu_k \cos\theta)\left(\frac{h}{\sin\theta}\right)$$

$$KE_f = mgh\left(1 - \frac{\mu_k}{\tan\theta}\right)$$

3. CALCULATE **Substitute the values into the equation(s) and solve:**

$$KE_f = (1.37 \times 10^4 \text{ kg})(9.81 \text{ m/s}^2)(147 \text{ m})\left(1 - \frac{0.45}{\tan 52.0°}\right)$$

$$KE_f = (1.37 \times 10^4 \text{ kg})(9.81 \text{ m/s}^2)(147 \text{ m})(1.00 - 0.35)$$

$$KE_f = (1.37 \times 10^4 \text{ kg})(9.81 \text{ m/s}^2)(147 \text{ m})(0.65)$$

$$KE_f = \boxed{1.3 \times 10^7 \text{ J}}$$

4. EVALUATE Note that the net force, and thus the final kinetic energy, is about two-thirds of what it would be if the side of the pyramid were frictionless.

ADDITIONAL PRACTICE

1. The tops of the towers of the Golden Gate Bridge, in San Francisco, are 227 m above the water. Suppose a worker drops a 655 g wrench from the top of a tower. If the average force of air resistance is 2.20 percent of the force of free fall, what will the kinetic energy of the wrench be when it hits the water?

2. Bonny Blair of the United States set a world record in speed skating when she skated 5.00×10^2 m with an average speed of 12.92 m/s. Suppose Blair crossed the finish line at this speed and then skated to a stop. If the work done by friction over a certain distance was −2830 J, what would Blair's kinetic energy be, assuming her mass to be 55.0 kg.

3. The CN Tower in Toronto, Canada, is 553 m tall, making it the tallest free-standing structure in the world. Suppose a chunk of ice with a mass of 25.0 g falls from the top of the tower. The speed of the ice is 30.0 m/s as it passes the restaurant, which is located 353 m above the ground. What is the magnitude of the average force due to air resistance?

4. In 1979, Dr. Hans Liebold of Germany drove a race car 12.6 km with an average speed of 404 km/h. Suppose Liebold applied the brakes to reduce his speed. What was the car's final speed if −3.00 MJ of work was done by the brakes? Assume the combined mass of the car and driver to be 1.00×10^3 kg.

5. The summit of Mount Everest is 8848.0 m above sea level, making it the highest summit on Earth. In 1953, Edmund Hillary was the first person to reach the summit. Suppose upon reaching there, Hillary slid a rock with a 45.0 g mass down the side of the mountain. If the rock's speed is

27.0 m/s when it is 8806.0 m above sea level, how much work was done on the rock by air resistance?

6. In 1990, Roger Hickey of California reached a speed 35.0 m/s on his skateboard. Suppose it took 21 kJ of work for Roger to reach this speed from a speed of 25.0 m/s. Calculate Hickey's mass.

7. At the 1984 Winter Olympics, William Johnson of the United States reached a speed of 104.5 km/h in the downhill skiing competition. Suppose Johnson left the slope at that speed and then slid freely along a horizontal surface. If the coefficient of kinetic friction between the skis and the snow was 0.120 and his final speed was half of his initial speed, find the distance William traveled.

Work and Energy

Problem D

POTENTIAL ENERGY

PROBLEM

In 1993, Javier Sotomayor from Cuba set a record in the high jump by clearing a vertical distance of 2.45 m. If the gravitational potential energy associated with Sotomayor at the top point of his trajectory was 1.59×10^3 J, what was his mass?

SOLUTION

Given: $h = 2.45$ m

$g = 9.81 \text{ m/s}^2$

$PE_g = 1.59 \times 10^3$ J

Unknown: $m = ?$

Use the equation for gravitational potential energy, and rearrange it to solve for *m*.

$$PE_g = mgh$$

$$m = \frac{PE_g}{gh}$$

$$m = \frac{(1.59 \times 10^3 \text{ J})}{\left(9.81 \frac{\text{m}}{\text{s}^2}\right)(2.45 \text{ m})} = \boxed{66.2 \text{ kg}}$$

ADDITIONAL PRACTICE

1. In 1992, Ukrainian Sergei Bubka used a short pole to jump to a height of 6.13 m. If the maximum potential energy associated with Bubka was 4.80 kJ at the midpoint of his jump, what was his mass?

2. Naim Suleimanoglu of Turkey has a mass of about 62 kg, yet he can lift nearly 3 times this mass. (This feat has earned Suleimanoglu the nickname of "Pocket Hercules.") If the potential energy associated with a barbell lifted 1.70 m above the floor by Suleimanoglu is 3.04×10^3 J, what is the barbell's mass?

3. In 1966, a special research cannon built in Arizona shot a projectile to a height of 180 km above Earth's surface. The potential energy associated with the projectile when its altitude was 10.0 percent of the maximum height was 1.48×10^7J. What was the projectile's mass? Assume that constant free-fall acceleration at this altitude is the same as at sea level.

4. The highest-caliber cannon ever built (though never used) is located in Moscow, Russia. The diameter of the cannon's barrel is about 89 cm, and the cannon's mass is 3.6×10^4 kg. Suppose this cannon were lifted by airplane. If the potential energy associated with this cannon were

8.88×10^8 J, what would be its height above sea level? Assume that constant free-fall acceleration at this altitude is the same as at sea level.

5. In 1987, Stefka Kostadinova from Bulgaria set a new women's record in high jump. It is known that the ratio of the potential energy associated with Kostadinova at the top of her jump to her mass was 20.482 m^2/s^2. What was the height of her record jump?

6. In 1992, David Engwall of California used a slingshot to launch a dart with a mass of 62 g. The dart traveled a horizontal distance of 477 m. Suppose the slingshot had a spring constant of 3.0×10^4 N/m. If the elastic potential energy stored in the slingshot just before the dart was launched was 1.4×10^2 J, how far was the slingshot stretched?

7. Suppose a 51 kg bungee jumper steps off the Royal Gorge Bridge, in Colorado. The bridge is situated 321 m above the Arkansas River. The bungee cord's spring constant is 32 N/m, the cord's relaxed length is 104 m, and its length is 179 m when the jumper stops falling. What is the total potential energy associated with the jumper at the end of his fall? Assume that the bungee cord has negligible mass.

8. Situated 4080 m above sea level, La Paz, Bolivia, is the highest capital in the world. If a car with a mass of 905 kg is driven to La Paz from a location that is 1860 m above sea level, what is the increase in potential energy?

9. In 1872, a huge gold nugget with a mass of 286 kg was discovered in Australia. The nugget was displayed for the public before it was melted down to extract pure gold. Suppose this nugget is attached to the ceiling by a spring with a spring constant of 9.50×10^3 N/m. The nugget is released from a height of 1.70 m above the floor, and is caught when it is no longer moving downward and is about to be pulled back up by the elastic force of the spring.

a. If the spring stretches a total amount of 59.0 cm, what is the elastic potential energy associated with the spring-nugget system?

b. What is the gravitational potential energy associated with the nugget just before it is dropped?

c. What is the gravitational potential energy associated with the nugget after the spring has stretched 59.0 cm?

d. What is the difference between the gravitational potential energy values in parts (b) and (c)? How does this compare with your answer for part (a)?

10. When April Moon set a record for flight shooting in 1981, the arrow traveled a distance of 9.50×10^2 m. Suppose the arrow had a mass of 65.0 g, and that the angle at which the arrow was launched was 45.0° above the horizontal.

a. What was the kinetic energy of the arrow at the instant it left the bowstring?

b. If the bowstring was pulled back 55.0 cm from its relaxed position, what was the spring constant of the bowstring? (Hint: Assume that all of the elastic potential energy stored in the bowstring is converted to the arrow's initial kinetic energy.)

c. Assuming that air resistance is negligible, determine the maximum height that the arrow reaches. (Hint: Equate the arrow's initial kinetic energy to the sum of the maximum gravitational potential energy associated with the arrow and the arrow's kinetic energy at maximum height.)

NAME ______________________ DATE ____________ CLASS ____________

Work and Energy

Problem E

CONSERVATION OF MECHANICAL ENERGY

PROBLEM

The largest apple ever grown had a mass of about 1.47 kg. Suppose you hold such an apple in your hand. You accidentally drop the apple, then manage to catch it just before it hits the ground. If the speed of the apple at that moment is 5.42 m/s, what is the kinetic energy of the apple? From what height did you drop it?

SOLUTION

1. DEFINE **Given:**

$m = 1.47$ kg

$g = 9.81$ m/s^2

$v = 5.42$ m/s

Unknown: $KE = ?$ $h = ?$

2. PLAN **Choose the equation(s) or situation:** Use the equations for kinetic and gravitational potential energy.

$$KE = \frac{1}{2}mv^2$$

$$PE_g = mgh$$

The zero level for gravitational potential energy is the ground. Because the apple ends its fall at the zero level, the final gravitational potential energy is zero.

$$PE_{g,f} = 0$$

The initial gravitational potential energy is measured at the point from which the apple is released.

$$PE_{g,i} = mgh$$

The apple is initially at rest, so the initial kinetic energy is zero.

$$KE_i = 0$$

The final kinetic energy is therefore:

$$KE_f = \frac{1}{2}mv^2$$

3. CALCULATE **Substitute the values into the equation(s) and solve:**

$$PE_{g,i} = (1.47 \text{ kg})(9.81 \frac{\text{m}}{\text{s}^2})h$$

$$KE_f = \frac{1}{2}(1.47 \text{ kg})\left(5.42 \frac{\text{m}}{\text{s}}\right)^2$$

Solving for KE yields the following result:

$$KE = KE_f = \frac{1}{2}(1.47 \text{ kg})\left(5.42 \frac{\text{m}}{\text{s}}\right)^2 = \boxed{21.6 \text{ J}}$$

Now use the principle of conservation for mechanical energy and the calculated quantity for KE_f to evaluate h.

$$ME_i = ME_f$$

$$PE_i + KE_i = PE_f + KE_f$$

$$PE_i + 0\ \text{J} = 0\ \text{J} + 21.6\ \text{J}$$

$$mgh = 21.6\ \text{J}$$

$$h = \frac{21.6\ \text{J}}{(1.47\ \text{kg})\left(9.81\ \frac{\text{m}}{\text{s}^2}\right)} = \boxed{1.50\ \text{m}}$$

4. EVALUATE Note that the height of the apple can be determined without knowing the apple's mass. This is because the conservation equation reduces to the equation relating speed, acceleration, and displacement: $v^2 = 2gh$.

ADDITIONAL PRACTICE

1. The largest watermelon ever grown had a mass of 118 kg. Suppose this watermelon were exhibited on a platform 5.00 m above the ground. After the exhibition, the watermelon is allowed to slide along to the ground along a smooth ramp. How high above the ground is the watermelon at the moment its kinetic energy is 4.61 kJ?

2. One species of eucalyptus tree, *Eucalyptus regnens*, grows to heights similar to those attained by California redwoods. Suppose a bird sitting on top of one specimen of eucalyptus tree drops a nut. If the speed of the falling nut at the moment it is 50.0 m above the ground is 42.7 m/s, how tall is the tree? Do you need to know the mass of the nut to solve this problem? Disregard air resistance.

3. In 1989, Michel Menin of France walked on a tightrope suspended under a balloon nearly at an altitude of 3150 m above the ground. Suppose a coin falls from Menin's pocket during his walk. How high above the ground is the coin when its speed is 60.0 m/s?

4. In 1936, Col. Harry Froboess of Switzerland jumped into the ocean from the airship *Graf Hindenburg*, which was 1.20×10^2 m above the water's surface. Assuming Froboess had a mass of 72.0 kg, what was his kinetic energy at the moment he was 30.0 m from the water's surface? What was his speed at that moment? Neglect the air resistance.

5. Suppose a motorcyclist rides a certain high-speed motorcycle. He reaches top speed and then coasts up a hill. The maximum height reached by the motorcyclist is 250.0 m. If 2.55×10^5 J of kinetic energy is dissipated by friction, what was the initial speed of the motorcycle? The combined mass of the motorcycle and motorcyclist is 250.0 kg.

6. The deepest mine ever drilled has a depth of 12.3 km (by contrast, Mount Everest has height of 8.8 km). Suppose you drop a rock with a mass of 120.0 g down the shaft of this mine. What would the rock's kinetic energy be after falling 3.2 km? What would the potential energy associated with the rock be at that same moment? Assume no air resistance and a constant free-fall acceleration.

7. *Desperado,* a roller coaster built in Nevada, has a vertical drop of 68.6 m. The roller coaster is designed so that the speed of the cars at the end of this drop is 35.6 m/s. Assume the cars are at rest at the start of the drop. What percent of the initial mechanical energy is dissipated by friction?

Work and Energy

Problem F

POWER

PROBLEM

Martinus Kuiper of the Netherlands ice skated for 24 h with an average speed of 6.3 m/s. Suppose Kuiper's mass was 65 kg. If Kuiper provided 520 W of power to accelerate for 2.5 s, how much work did he do?

SOLUTION

Given: $P = 520$ W

$\Delta t = 2.5$ s

Unknown: $W = ?$

Use the equation for power and rearrange it to solve for work.

$$P = \frac{W}{\Delta t}$$

$$W = P\Delta t = (520 \text{ W})(2.5 \text{ s}) = \boxed{1300 \text{ J}}$$

ADDITIONAL PRACTICE

1. The most powerful ice breaker in the world was built in the former Soviet Union. The ship is almost 150 m long, and its nuclear engine generates 56 MW of power. How much work can this engine do in 1.0 h?
2. Reginald Esuke from Cameroon ran over 3 km down a mountain slope in just 62.25 min. How much work was done if the power developed during Esuke's descent was 585.0 W?
3. The world's tallest lighthouse is located in Japan and is 106 m tall. A winch that provides 3.00×10^2 W of power is used to raise 14.0 kg of equipment to the lighthouse top at a constant velocity. How long does it take the equipment to reach the lighthouse top?
4. The first practical car to use a gasoline engine was built in London in 1826. The power generated by the engine was just 2984 W. How long would this engine have to run to produce 3.60×10^4 J of work?
5. Dennis Joyce of the United States threw a boomerang and caught it at the same location 3.0 min later. Suppose Joyce decided to work out while waiting for the boomerang to return. If he expended 54 kJ of work, what was his average power output during the workout?
6. In 1984, Don Cain threw a flying disk that stayed aloft for 16.7 s. Suppose Cain ran up a staircase during this time, reaching a height of 18.4 m. If his mass was 72.0 kg, how much power was needed for Cain's ascent.

Momentum and Collisions

Problem A

MOMENTUM

PROBLEM

The world's most massive train ran in South Africa in 1989. Over 7 km long, the train traveled 861.0 km in 22.67 h. Imagine that the distance was traveled in a straight line north. If the train's average momentum was 7.32×10^8 kg•m/s to the north, what was its mass?

SOLUTION

Given: $\Delta \mathbf{x} = 861.0$ km to the north

$\Delta t = 22.67$ h

$\mathbf{p_{avg}} = 7.32 \times 10^8 \frac{\text{kg•m}}{\text{s}}$ to the north

Unknown: $\mathbf{v_{avg}} = ?$ $m = ?$

Use the definition of average velocity to calculate $\mathbf{v_{avg}}$, and then substitute this value for velocity in the definition of momentum to solve for mass.

$$\mathbf{v_{avg}} = \frac{\Delta \mathbf{x}}{\Delta t} = \frac{(861.0 \times 10^3 \text{ m})}{(22.67 \text{ h})(3600 \text{ s/h})} = 10.55 \frac{\text{m}}{\text{s}} \text{ to the north}$$

$$p_{avg} = mv_{avg}$$

$$m = \frac{p_{avg}}{v_{avg}} = \frac{\left(7.32 \times 10^8 \frac{\text{kg•m}}{\text{s}}\right)}{\left(10.55 \frac{\text{m}}{\text{s}}\right)} = \boxed{6.94 \times 10^7 \text{ kg}}$$

ADDITIONAL PRACTICE

1. In 1987, Marisa Canofoglia, of Italy, roller-skated at a record-setting speed of 40.3 km/h. If the magnitude of Canofoglia's momentum was 6.60×10^2 kg•m/s, what was her mass?

2. In 1976, a 53 kg helicopter was built in Denmark. Suppose this helicopter flew east with a speed of 60.0 m/s and the total momentum of the helicopter and pilot was 7.20×10^3 kg•m/s to the east. What was the mass of the pilot?

3. One of the smallest planes ever flown was the *Bumble Bee II,* which had a mass of 1.80×10^2 kg. If the pilot's mass was 7.0×10^1 kg, what was the velocity of both plane and pilot if their momentum was 2.08×10^4 kg•m/s to the west?

4. The first human-made satellite, *Sputnik I,* had a mass of 83.6 kg and a momentum with a magnitude of 6.63×10^5 kg•m/s. What was the satellite's speed?

5. Among the largest passenger ships currently in use, the *Norway* has been in service the longest. The *Norway* is more than 300 m long, has a mass of 6.9×10^7 kg, and can reach a top cruising speed of 33 km/h. Calculate the magnitude of the ship's momentum.

6. In 1994, a tower 22.13 m tall was built of Lego® blocks. Suppose a block with a mass of 2.00 g is dropped from the top of this tower. Neglecting air resistance, calculate the block's momentum at the instant the block hits the ground.

Momentum and Collisions

Problem B

FORCE AND IMPULSE

PROBLEM

In 1993, a generator with a mass of 1.24×10^5 kg was flown from Germany to a power plant in India on a Ukrainian-built plane. This constituted the heaviest single piece of cargo ever carried by a plane. Suppose the plane took off with a speed of 101 m/s toward the southeast and then accelerated to a final cruising speed of 197 m/s. During this acceleration, a force of 4.00×10^5 N in the southeast direction was exerted on the generator. For how much time did the force act on the generator?

SOLUTION

Given: $m = 1.24 \times 10^5$ kg

$\mathbf{v_i} = 101$ m/s to the southeast

$\mathbf{v_f} = 197$ m/s to the southeast

$\mathbf{F} = 4.00 \times 10^5$ N to the southeast

Unknown: $\Delta t = ?$

Use the impulse-momentum theorem to determine the time the force acts on the generator.

$$\mathbf{F}\Delta t = \Delta \mathbf{p} = m\mathbf{v_f} - m\mathbf{v_i}$$

$$\Delta t = \frac{m\mathbf{v_f} - m\mathbf{v_i}}{\mathbf{F}}$$

$$\Delta t = \frac{(1.24 \times 10^5 \text{ kg})(197 \text{ m/s}) - (1.24 \times 10^5 \text{ kg})(101 \text{ m/s})}{4.00 \times 10^5 \text{ N}}$$

$$\Delta t = \frac{2.44 \times 10^7 \text{ kg•m/s} - 1.25 \times 10^7 \text{ kg•m/s}}{4.00 \times 10^5 \text{ N}}$$

$$\Delta t = \frac{1.19 \times 10^7 \text{ kg•m/s}}{4.00 \times 10^5 \text{ N}}$$

$$\Delta t = \boxed{29.8 \text{ s}}$$

ADDITIONAL PRACTICE

1. In 1991, a Swedish company, Kalmar LMV, constructed a forklift truck capable of raising 9.0×10^4 kg to a height of about 2 m. Suppose a mass this size is lifted with an upward velocity of 12 cm/s. The mass is initially at rest and reaches its upward speed because of a net force of 6.0×10^3 N exerted upward. For how long is this force applied?

2. A bronze statue of Buddha was completed in Tokyo in 1993. The statue is 35 m tall and has a mass of 1.00×10^6 kg. Suppose this statue were to be moved to a different location. What is the magnitude of the impulse that must act on the statue in order for the speed to increase from 0 m/s

to 0.20 m/s? If the magnitude of the net force acting on the statue is 12.5 kN, how long will it take for the final speed to be reached?

3. In 1990, Gary Stewart of California made 177 737 jumps on a pogo stick. Suppose that the pogo stick reaches a height of 12.0 cm with each jump and that the average net force acting on the pogo stick during the contact with the ground is 330 N upward. What is the time of contact with the ground between the jumps? Assume the total mass of Stewart and the pogo stick is 65 kg. (Hint: The difference between the initial and final velocities is one of direction rather than magnitude.)

4. The specially designed armored car that was built for Leonid Brezhnev when he was head of the Soviet Union had a mass of about 6.0×10^3 kg. Suppose this car is accelerated from rest by a force of 8.0 kN to the east. What is the car's velocity after 8.0 s?

5. In 1992, Dan Bozich of the United States drove a gasoline-powered go-cart at a speed of 125.5 km/h. Suppose Bozich applies the brakes upon reaching this speed. If the combined mass of the go-cart and driver is 2.00×10^2 kg, the decelerating force is 3.60×10^2 N opposite the cart's motion, and the time during which the deceleration takes place is 10.0 s. What is the final speed of Bozich and the go-cart?

6. The "human cannonball" has long been a popular—and extremely dangerous—circus stunt. In order for a 45 kg person to leave the cannon with the fastest speed yet achieved by a human cannonball, a 1.6×10^3 N force must be exerted on that person for 0.68 s. What is the record speed at which a person has been shot from a circus cannon?

7. The largest steam-powered locomotive was built in the United States in 1943. It is still operational and is used for entertainment purposes. The locomotive's mass is 4.85×10^5 kg. Suppose this locomotive is traveling northwest along a straight track at a speed of 20.0 m/s. What force must the locomotive exert to increase its velocity to 25.0 m/s to the northwest in 5.00 s?

8. With upward speeds of 12.5 m/s, the elevators in the Yokohama Landmark Tower in Yokohama, Japan, are among the fastest elevators in the world. Suppose a passenger with a mass of 70.0 kg enters one of these elevators. The elevator then goes up, reaching full speed in 4.00 s. Calculate the net force that is applied to the passenger during the elevator's acceleration.

9. Certain earthworms living in South Africa have lengths as great as 6.0 m and masses as great as 12.0 kg. Suppose an eagle picks up an earthworm of this size, only to drop it after both have reached a height of 40.0 m above the ground. By skillfully using its muscles, the earthworm manages to extend the time during which it collides with the ground to 0.250 s. What is the net force that acts on the earthworm during its collision with the ground? Assume the earthworm's vertical speed when it is initially dropped to be 0 m/s.

Momentum and Collisions

Problem C

STOPPING DISTANCE

PROBLEM

The largest nuts (and, presumably, the largest bolts) are manufactured in England. The nuts have a mass of 4.74 × 10^3 kg each, which is greater than any passenger car currently in production. Suppose one of these nuts slides along a rough horizontal surface with an initial velocity of 2.40 m/s to the right. If the force of friction acting on the nut is 6.8 × 10^3 N to the left, what is the change in the nut's momentum after 1.1 s? How far does the nut travel during its change in momentum?

SOLUTION

Given:
$m = 4.74 \times 10^3$ kg
$\mathbf{v_i} = 2.40$ m/s to the right $= +2.40$ m/s
$\mathbf{F} = 6.8 \times 10^3$ N to the left $= -6.8 \times 10^3$ N
$\Delta t = 1.1$ s

Unknown: $\Delta\mathbf{p} = ?$ $\quad \Delta\mathbf{x} = ?$

Use the impulse-momentum theorem to calculate the change in momentum. Use the definition of momentum to find $\mathbf{v_f}$, and then use the equation for stopping distance to solve for $\Delta\mathbf{x}$.

$$\Delta\mathbf{p} = \mathbf{F}\Delta\mathbf{t} = (-6.8 \times 10^3 \text{ N})(1.1 \text{ s})$$

$$\Delta\mathbf{p} = \boxed{-7.5 \times 10^3 \text{ kg•m/s to the right, or } 7.5 \times 10^3 \text{ kg•m/s to the left}}$$

$$\Delta\mathbf{p} = m\mathbf{v_f} - m\mathbf{v_i}$$

$$\mathbf{v_f} = \frac{\Delta\mathbf{p} + m\mathbf{v_i}}{m}$$

$$\mathbf{v_f} = \frac{-7.5 \times 10^3 \text{ kg•m/s} + (4.74 \times 10^3 \text{ kg})(2.40 \text{ m/s})}{4.74 \times 10^3 \text{ kg}}$$

$$\mathbf{v_f} = \frac{-7.5 \times 10^3 \text{ kg•m/s} + 1.14 \times 10^4 \text{ kg•m/s}}{4.74 \times 10^3 \text{ kg}} = \frac{3900 \text{ kg•m/s}}{4.74 \times 10^3 \text{ kg}}$$

$$\mathbf{v_f} = \boxed{0.82 \text{ m/s to the right}}$$

$$\Delta\mathbf{x} = \tfrac{1}{2}(\mathbf{v_i} + \mathbf{v_f})\Delta t = \tfrac{1}{2}(2.40 \text{ m/s} + 0.82 \text{ m/s})(1.1 \text{ s}) = \tfrac{1}{2}(3.22 \text{ m/s})(1.1 \text{ s})$$

$$\Delta\mathbf{x} = \boxed{1.8 \text{ m/s to the right}}$$

ADDITIONAL PRACTICE

1. The most powerful tugboats in the world are built in Finland. These boats exert a force with a magnitude of 2.85 × 10^6 N. Suppose one of these tugboats is trying to slow a huge barge that has a mass of 2.0 × 10^7 kg and is moving with a speed of 3.0 m/s. If the tugboat exerts its maximum force for 21 s in the direction opposite to that in which the barge is moving, what will be the change in the barge's momentum? How far will the barge travel before it is brought to a stop?

2. In 1920, a 6.5×10^4 kg meteorite was found in Africa. Suppose a meteorite with this mass enters Earth's atmosphere with a speed of 1.0 km/s. What is the change in the meteorite's momentum if an average constant force of -1.7×10^6 N acts on the meteorite for 30.0 s? How far does the meteorite travel during this time?

3. The longest canoe in the world was constructed in New Zealand. The combined mass of the canoe and its crew of more than 70 people was 2.03×10^4 kg. Suppose the canoe is rowed from rest to a velocity of 5.00 m/s to the east, at which point the crew takes a break for 20.3 s. If a constant average retarding force of 1.20×10^3 N to the west acts on the canoe, what is the change in the momentum of the canoe and crew? How far does the canoe travel during the time the crew is not rowing?

4. The record for the smallest dog in the world belongs to a terrier who had a mass of only 113 g. Suppose this dog runs to the right with a speed of 2.00 m/s when it suddenly sees a mouse. The dog becomes scared and uses its paws to bring itself to rest in 0.80 s. What is the force required to stop the dog? What is the dog's stopping distance?

5. In 1992, an ice palace estimated to be 4.90×10^6 kg was built in Minnesota. Despite this sizable mass, this structure could be moved at a constant velocity because of the small force of friction between the ice blocks of its base and the ice of the lake upon which it was constructed. Imagine moving the entire palace with a speed of 0.200 m/s on this very smooth, icy surface. Once the palace is no longer being pushed, it coasts to a stop in 10.0 s. What is the average force of kinetic friction acting on the palace? What is the palace's stopping distance?

6. *Steel Phantom* is a roller coaster in Pennsylvania that, like the *Desperado* in Nevada, has a vertical drop of 68.6 m. Suppose a roller-coaster car with a mass of 1.00×10^3 kg travels from the top of that drop without friction. The car then decelerates along a horizontal stretch of track until it comes to a stop. How long does it take the car to decelerate if the constant force acting on it is -2.24×10^4 N? How far does the car travel along the horizontal track before stopping? Assume the car's speed at the peak of the drop is zero.

7. Two Japanese islands are connected by a long rail tunnel that extends horizontally underwater. Imagine a communication system in which a small rail car with a mass of 100.0 kg is launched by a type of cannon in order to transport messages between the two islands. Assume a rail car from one end of the tunnel has a speed of 4.50×10^2 m/s, which is just large enough for a constant frictional force of -188 N to cause the car to stop at the other end of the tunnel. How long does it take for the car to travel the length of the tunnel? What is the length of the tunnel?

NAME ______________________ DATE ____________ CLASS ______________

Momentum and Collisions

Problem D

CONSERVATION OF MOMENTUM

PROBLEM

Kangaroos are good runners that can sustain a speed of 56 km/h (15.5 m/s). Suppose a kangaroo is sitting on a log that is floating in a lake. When the kangaroo gets scared, she jumps off the log with a velocity of 15 m/s toward the bank. The log moves with a velocity of 3.8 m/s away from the bank. If the mass of the log is 250 kg, what is the mass of the kangaroo?

SOLUTION

1. DEFINE **Given:**

$\mathbf{v_{k,i}}$ = initial velocity of kangaroo = 0 m/s

$\mathbf{v_{l,i}}$ = initial velocity of log = 0 m/s

$\mathbf{v_{k,f}}$ = final velocity of kangaroo = 15 m/s toward bank = +15 m/s

$\mathbf{v_{l,f}}$ = final velocity of log = 3.8 m/s away from bank = −3.8 m/s

m_l = mass of log = 250 kg

Unknown: m_k = mass of kangaroo = ?

2. PLAN **Choose the equation(s) or situation:** Because the momentum of the kangaroo-log system is conserved and therefore remains constant, the total initial momentum of the kangaroo and log will equal the total final momentum of the kangaroo and log.

$$m_k\mathbf{v_{k,i}} + m_l\mathbf{v_{l,i}} = m_k\mathbf{v_{k,f}} + m_l\mathbf{v_{l,f}}$$

The initial velocities of the kangaroo and log are zero, and therefore the initial momentum for each of them is zero. It thus follows that the total final momentum for the kangaroo and log must also equal zero. The momentum-conservation equation reduces to the following:

$$m_k\mathbf{v_{k,f}} + m_l\mathbf{v_{l,f}} = 0$$

Rearrange the equation(s) and isolate the unknown(s):

$$m_k = \frac{-m_l\mathbf{v_{l,f}}}{\mathbf{v_{k,f}}}$$

3. CALCULATE **Substitute** the values into the equation(s) and solve:

$$m_k = \frac{-(250\ \text{kg})(-3.8\ \text{m/s})}{15\ \text{m/s}}$$

$$m_k = \boxed{63\ \text{kg}}$$

4. EVALUATE Because the log is about four times as massive as the kangaroo, its velocity is about one-fourth as large as the kangaroo's velocity.

NAME ______________________ DATE ______________ CLASS ______________

ADDITIONAL PRACTICE

1. The largest single publication in the world is the 1112-volume set of *British Parliamentary Papers* for 1968 through 1972. The complete set has a mass of 3.3×10^3 kg. Suppose the entire publication is placed on a cart that can move without friction. The cart is at rest, and a librarian is sitting on top of it, just having loaded the last volume. The librarian jumps off the cart with a horizontal velocity relative to the floor of 2.5 m/s to the right. The cart begins to roll to the left at a speed of 0.05 m/s. Assuming the cart's mass is negligible, what is the librarian's mass?

2. The largest grand piano in the world is *really* grand. Built in London, it has a mass of 1.25×10^3 kg. Suppose a pianist finishes playing this piano and pushes herself from the piano so that she rolls backwards with a speed of 1.4 m/s. Meanwhile, the piano rolls forward so that in 4.0 s it travels 24 cm at constant velocity. Assuming the stool that the pianist is sitting on has a negligible mass, what is the pianist's mass?

3. With a mass of 114 kg, *Baby Bird* is the smallest monoplane ever flown. Suppose the *Baby Bird* and pilot are coasting along the runway when the pilot jumps horizontally to the runway behind the plane. The pilot's velocity upon leaving the plane is 5.32 m/s backward. After the pilot jumps from the plane, the plane coasts forward with a speed of 3.40 m/s. If the pilot's mass equals 60.0 kg, what is the velocity of the plane and pilot before the pilot jumps?

4. The September 14, 1987, issue of the *New York Times* had a mass of 5.4 kg. Suppose a skateboarder picks up a copy of this issue to have a look at the comic pages while rolling backward on the skateboard. Upon realizing that the *New York Times* doesn't have a "funnies" section, the skateboarder promptly throws the entire issue in a recycling container. The newspaper is thrown forward with a speed of 7.4 m/s. When the skater throws the newspaper away, he rolls backward at a speed of 1.4 m/s. If the combined mass of the skateboarder and skateboard is assumed to be 50.0 kg, what is the initial velocity of the skateboarder and newspaper?

5. The longest bicycle in the world was built in New Zealand in 1988. It is more than 20 m in length, has a mass of 3.4×10^2 kg, and can be ridden by four people at a time. Suppose four people are riding the bike southeast when they realize that the street turns and that the bike won't make it around the corner. All four riders jump off the bike at the same time and with the same velocity (9.0 km/h to the northwest, as measured relative to Earth). The bicycle continues to travel forward with a velocity of 28 km/h to the southeast. If the combined mass of the riders is 2.6×10^2 kg, what is the velocity of the bicycle and riders immediately before the riders' escape?

6. The largest frog ever found was discovered in Cameroon in 1989. The frog's mass was nearly 3.6 kg. Suppose this frog is placed on a skateboard with a mass of 3.0 kg. The frog jumps horizontally off the skateboard to the right, and the skateboard rolls freely in the opposite direction with a speed of 2.0 m/s relative to the ground. If the frog and skateboard are initially at rest, what is the initial horizontal velocity of the frog?

7. In 1994, a pumpkin with a mass of 449 kg was grown in Canada. Suppose you want to push a pumpkin with this mass along a smooth, horizontal ramp. You give the pumpkin a good push, only to find yourself sliding backwards at a speed of 4.0 m/s. How far will the pumpkin slide 3.0 s after the push? Assume your mass to be 60.0 kg.

Momentum and Collisions

Problem E

PERFECTLY INELASTIC COLLISIONS

PROBLEM

The Chinese giant salamander is one of the largest of salamanders. Suppose a Chinese giant salamander chases a 5.00 kg carp with a velocity of 3.60 m/s to the right and the carp moves with a velocity of 2.20 m/s in the same direction (away from the salamander). If the speed of the salamander and carp immediately after the salamander catches the carp is 3.50 m/s to the right, what is the salamander's mass?

SOLUTION

Given: m_c = mass of carp = 5.00 kg

$\mathbf{v_{s,i}}$ = initial velocity of salamander = 3.60 m/s to the right

$\mathbf{v_{c,i}}$ = initial velocity of carp = 2.20 m/s to the right

$\mathbf{v_f}$ = final velocity = 3.50 m/s to the right

Unknown: m_s = mass of salamander = ?

Use the equation for a perfectly inelastic collision and rearrange it to solve for m_s.

$$m_s\mathbf{v_{s,i}} + m_c\mathbf{v_{c,i}} = (m_s + m_c)\mathbf{v_f}$$

$$m_s = \frac{m_c\mathbf{v_f} - m_c\mathbf{v_{c,i}}}{\mathbf{v_{s,i}} - \mathbf{v_f}}$$

$$m_s = \frac{(5.00 \text{ kg})(3.50 \text{ m/s}) - (5.00 \text{ kg})(2.20 \text{ m/s})}{3.60 \text{ m/s} - 3.50 \text{ m/s}}$$

$$m_s = \frac{17.5 \text{ kg}\bullet\text{m/s} - 11.0 \text{ kg}\bullet\text{m/s}}{0.10 \text{ m/s}}$$

$$m_s = \frac{6.5 \text{ kg}\bullet\text{m/s}}{0.10 \text{ m/s}}$$

$$m_s = \boxed{65 \text{ kg}}$$

ADDITIONAL PRACTICE

1. Zorba, an English mastiff with a mass of 155 kg, jumps forward horizontally at a speed of 6.0 m/s into a boat that is floating at rest. After the jump, the boat and Zorba move with a velocity of 2.2 m/s forward. Calculate the boat's mass.

2. Yvonne van Gennip of the Netherlands ice skated 10.0 km with an average speed of 10.8 m/s. Suppose van Gennip crosses the finish line at her average speed and takes a huge bouquet of flowers handed to her by a fan. As a result, her speed drops to 10.01 m/s. If van Gennip's mass is 63.0 kg, what is the mass of the bouquet?

3. The world's largest guitar was built by a group of high school students in Indiana. Suppose that this guitar is placed on a light cart. The cart and guitar are then pushed with a velocity of 4.48 m/s to the right. One of the students tries to slow the cart by stepping on it as it passes by her. The new velocity of the cart, guitar, and student is 4.00 m/s to the right. If the student's mass is 54 kg, what is the mass of the guitar? Assume the mass of the cart is negligible.

4. The longest passenger buses in the world operate in Zaire. These buses are more than 30 m long, have two trailers, and have a total mass of 28×10^3 kg. Imagine a safety test involving one of these buses and a truck with a mass of 12×10^3 kg. The truck with an unknown velocity hits a bus that is at rest so that the two vehicles move forward together with a speed of 3.0 m/s. Calculate the truck's velocity prior to the collision.

5. Sumo wrestlers must be very heavy to be successful in their sport, which involves pushing the rival out of the ring. One of the greatest sumo champions, Akebono, had a mass of 227 kg. The heaviest sumo wrestler ever, Konishiki, at one point had a mass of 267 kg. Suppose these two wrestlers are opponents in a match. Akebono moves left with a speed of 4.0 m/s, while Konishiki moves toward Akebono with an unknown speed. After the wrestlers undergo an inelastic collision, both have a velocity of zero. From this information, calculate Konishiki's velocity before colliding with Akebono.

6. Louis Borsi, of London, built a drivable car that had a mass of 9.50 kg and could move as fast as 24.0 km/h. Suppose the inventor falls out of this car and the car proceeds driverless to the north at its maximum speed. The inventor's young daughter, who has a mass of 32.0 kg, "catches" the car by jumping northward from a nearby stairway. The velocity of the car and girl is 11.0 km/h to the north. What was the velocity of Borsi's daughter as she jumped in the car?

7. In 1990, Roger Hickey of California attained a speed of 89 km/h while standing on a skateboard. Suppose Hickey is riding horizontally at his stand-up speed when he catches up to another skateboarder, who is moving at 69 km/h in the same direction. If the second skateboarder steps sideways onto Hickey's skateboard, the two riders move forward with a new speed. Calculate this speed, assuming that both skateboarders have equal, but unknown, masses and that the mass of the skateboard is negligible.

8. The white shark is the largest carnivorous fish in the world. The mass of a white shark can be as great as 3.0×10^3 kg. In spite of (or perhaps because of) the mass and ferocity of the shark, it is prized by commercial and sports fishers alike. Suppose Joe, who is one of these fishers, goes to a cliff that overlooks the ocean. To see if the sharks are biting, Joe drops a 2.5×10^2 kg fish as bait into the ocean below. As it so happens, a 3.0×10^3 kg white shark is prowling the ocean floor just below the cliff. The shark sees the bait, which is sinking straight down at a speed of 3.0 m/s. The shark swims upward with a speed of 1.0 m/s to swallow the bait. What is the velocity of the shark right after it has swallowed the bait?

9. The heaviest cow on record had a mass of 2.267×10^3 kg and lived in Maine at the beginning of the twentieth century. Imagine that during an agricultural exhibition, the cow's owner puts the cow on a railed cart that has a mass of 5.00×10^2 kg and pushes the cow and cart left to the stage with a speed of 2.00 m/s. Another farmer puts his cow, which has a mass of 1.800×10^3 kg, on an identical cart and pushes it toward the stage from the opposite direction with a speed of 1.40 m/s. The carts collide and stick together. What is the velocity of the cows after the collision?

NAME ______________________ DATE ____________ CLASS ____________

Momentum and Collisions

Problem F

KINETIC ENERGY IN PERFECTLY INELASTIC COLLISIONS

PROBLEM

Alaskan moose can be as massive as 8.00×10^2 kg. Suppose two feuding moose, both of which have a mass of 8.00×10^2 kg, back away and then run toward each other. If one of them runs to the right with a speed of 8.0 m/s and the other runs to the left with a speed of 6.0 m/s, what amount of kinetic energy will be dissipated after their inelastic collision?

SOLUTION

Given:

m_1 = mass of first moose = 8.00×10^2 kg

m_2 = mass of second moose = 8.00×10^2 kg

$\mathbf{v_{1,i}}$ = initial velocity of first moose = 8.0 m/s to the right

$\mathbf{v_{2,i}}$ = initial velocity of second moose = 6.0 m/s to the left = −6.0 m/s to the right

Unknown: $\Delta KE = ?$

Use the equation for a perfectly inelastic collision.

$$m_1\mathbf{v_{1,i}} + m_2\mathbf{v_{2,i}} = (m_1 + m_2)\mathbf{v_f}$$

$$(8.00 \times 10^2\ \text{kg})(8.0\ \text{m/s}) + (8.00 \times 10^2\ \text{kg})(-6.0\ \text{m/s}) = (2)(8.00 \times 10^2\ \text{kg})\mathbf{v_f}$$

$$6.4 \times 10^3\ \text{kg}\bullet\text{m/s} - 4.8 \times 10^3\ \text{kg}\bullet\text{m/s} = (16.0 \times 10^2\ \text{kg})\mathbf{v_f}$$

$$\mathbf{v_f} = \frac{1.6 \times 10^3\ \text{kg}\bullet\text{m/s}}{16.0 \times 10^2\ \text{kg}} = 1.0\ \text{m/s to the right}$$

Use the equation for kinetic energy to calculate the kinetic energy of each moose before the collision and the final kinetic energy of the two moose combined.

Initial kinetic energy:

$$KE_i = KE_{1,i} + KE_{2,i} = \tfrac{1}{2}m_1v_{1,i}^2 + \tfrac{1}{2}m_2v_{2,i}^2$$

$$KE_i = \tfrac{1}{2}(8.00 \times 10^2\ \text{kg})(8.0\ \text{m/s})^2 + \tfrac{1}{2}(8.00 \times 10^2\ \text{kg})(-6.0\ \text{m/s})^2$$

$$KE_i = 2.6 \times 10^4\ \text{J} + 1.4 \times 10^4\ \text{J} = 4.0 \times 10^4\ \text{J}$$

Final kinetic energy:

$$KE_f = KE_{1,f} + KE_{2,f} = \tfrac{1}{2}(m_1 + m_2)v_f^2$$

$$KE_f = \frac{(2)(8.00 \times 10^2\ \text{kg})(1.0\ \text{m/s})^2}{2}$$

$$KE_f = 8.0 \times 10^2\ \text{J}$$

Change in kinetic energy:

$$\Delta KE = KE_f - KE_i = 8.0 \times 10^2\ \text{J} - 4.0 \times 10^4\ \text{J} = \boxed{-3.9 \times 10^4\ \text{J}}$$

By expressing ΔKE as a negative number, we indicate that the energy has left the system to take a form other than mechanical energy.

1. The hog-nosed bat is the smallest mammal on Earth: it is about the same size as a bumblebee and has an average mass of 2.0 g. Suppose a hog-nosed bat with this mass flies at 2.0 m/s when it detects a bug with a mass of 0.20 g flying directly toward it at 8.0 m/s. What fraction of the total kinetic energy is dissipated when it swallows the bug?

2. The heaviest wild lion ever measured had a mass of 313 kg. Suppose this lion is walking by a lake when it sees an empty boat floating at rest near the shore. The curious lion jumps into the boat with a speed of 6.00 m/s, causing the boat with the lion in it to move away from the shore with a speed of 2.50 m/s. How much kinetic energy is dissipated in this inelastic collision.

3. The cheapest car ever commercially produced was the *Red Bug Backboard*, which sold in 1922 in the United States for about $250. The car's mass was only 111 kg. Suppose two of these cars are used in a stunt crash for an action film. If one car's initial velocity is 9.00 m/s to the right and the other car's velocity is 5.00 m/s to the left, how much kinetic energy is dissipated in the crash?

4. In 1986, four high school students built an electric car that could reach a speed of 106.0 km/h. The mass of the car was just 60.0 kg. Imagine two of these cars used in a stunt show. One car travels east with a speed of 106.0 km/h, and the other car travels west with a speed of 75.0 km/h. If each car's driver has a mass of 50.0 kg, how much kinetic energy is dissipated in the perfectly inelastic head-on collision?

5. The Arctic Snow Train, built for the U.S. Army, has a mass of 4.00×10^5 kg and a top speed of 32.0 km/h. Suppose such a train moving at its top speed is hit from behind by another snow train with a mass of 1.60×10^5 kg and a speed of 45.0 km/h in the same direction. What is the change in kinetic energy after the trains' perfect inelastic collision?

6. There was a domestic cat in Australia with a mass of 21.3 kg. Suppose this cat is sitting on a skateboard that is not moving. A 1.80×10^{-1} kg treat is thrown to the cat. When the cat catches the treat, the cat and skateboard move with a speed of 6.00×10^{-2} m/s. How much kinetic energy is dissipated in the process? Assume one-dimensional motion.

7. In 1985, a spider with a mass of 122 g was caught in Surinam, South America. (Recall that the smallest dog in the world had a smaller mass.) Suppose a spider with this mass runs at a certain unknown speed when it collides inelastically with another spider, which has a mass of 96.0 g and is at rest. Find the fraction of the kinetic energy that is dissipated in the perfect inelastic collision. Assume that the resting spider is on a low-friction surface. Do you need to know the first spider's velocity to calculate the fraction of the dissipated kinetic energy?

NAME ______________________ DATE ______________ CLASS ______________

Momentum and Collisions

Problem G

ELASTIC COLLISIONS

PROBLEM

American juggler Bruce Sarafian juggled 11 identical balls at one time in 1992. Each ball had a mass of 0.20 kg. Suppose two balls have an elastic head-on collision during the act. The first ball moves away from the collision with a velocity of 3.0 m/s to the right, and the second ball moves away with a velocity of 4.0 m/s to the left. If the first ball's velocity before the collision is 4.0 m/s to the left, what is the velocity of the second ball before the collision?

SOLUTION

1. DEFINE **Given:** $m_1 = m_2 = 0.20$ kg

$\mathbf{v_{1,i}}$ = initial velocity of ball 1 = 4.0 m/s to the left
= −4.0 m/s to the right

$\mathbf{v_{1,f}}$ = final velocity of ball 1 = 3.0 m/s to the right

$\mathbf{v_{2,f}}$ = final velocity of ball 2 = 4.0 m/s to the left
= −4.0 m/s to the right

Unknown: $\mathbf{v_{2,i}}$ = initial velocity of ball 2 = ?

2. PLAN **Choose the equation(s) or situation:** Use the equation for the conservation of momentum to determine the initial velocity of ball 2. Because both balls have identical masses, the mass terms cancel.

$$m_1\mathbf{v_{1,i}} + m_2\mathbf{v_{2,i}} = m_1\mathbf{v_{1,f}} + m_2\mathbf{v_{2,f}}$$

$$\mathbf{v_{1,i}} + \mathbf{v_{2,i}} = \mathbf{v_{1,f}} + \mathbf{v_{2,f}}$$

Rearrange the equation(s) to isolate the unknown(s):

$$\mathbf{v_{2,i}} = \mathbf{v_{1,f}} + \mathbf{v_{2,f}} - \mathbf{v_{1,i}}$$

3. CALCULATE **Substitute the values into the equation(s) and solve:**

$$\mathbf{v_{2,i}} = 3.0 \text{ m/s} - 4.0 \text{ m/s} - (-4.0 \text{ m/s})$$

$$\mathbf{v_{2,i}} = \boxed{3.0 \text{ m/s to the right}}$$

4. EVALUATE Confirm your answer by making sure that kinetic energy is also conserved.

$$\frac{1}{2}m_1v_{1,i}^2 + \frac{1}{2}m_2v_{2,i}^2 = \frac{1}{2}m_1v_{1,f}^2 + \frac{1}{2}m_2v_{2,f}^2$$

$$v_{1,i}^2 + v_{2,i}^2 = v_{1,f}^2 + v_{2,f}^2$$

$$(-4.0 \text{ m/s})^2 + (3.0 \text{ m/s})^2 = (3.0 \text{ m/s})^2 + (-4.0 \text{ m/s})^2$$

$$16 \text{ m}^2/\text{s}^2 + 9.0 \text{ m}^2/\text{s}^2 = 9.0 \text{ m}^2/\text{s}^2 + 16 \text{ m}^2/\text{s}^2$$

$$25 \text{ m}^2/\text{s}^2 = 25 \text{ m}^2/\text{s}^2$$

ADDITIONAL PRACTICE

1. The moon's orbital speed around Earth is 3.680×10^3 km/h. Suppose the moon suffers a perfectly elastic collision with a comet whose mass is 50.0 percent that of the moon. (A partially inelastic collision would be a much

more realistic event.) After the collision, the moon moves with a speed of -4.40×10^2 km/h, while the comet moves away from the moon at 5.740×10^3 km/h. What is the comet's speed before the collision?

2. The largest beet root on record had a mass of 18.40 kg. The largest cabbage on record had a mass of 56.20 kg. Imagine these two vegetables traveling in opposite directions. The cabbage, which travels 5.000 m/s to the left, collides with the beet root. After the collision, the cabbage has a velocity of 6.600×10^{-2} m/s to the left, and the beet root has a velocity of 10.07 m/s to the left. What is the beet root's velocity before the perfectly elastic collision?

3. The first astronaut to walk in outer space without being tethered to a spaceship was Capt. Bruce McCandless. In 1984, he used a jet backpack, which cost about $15 million to design, to move freely about the exterior of the space shuttle *Challenger.* Imagine two astronauts working in outer space. Suppose they have equal masses and accidentally run into each other. The first astronaut moves 5.0 m/s to the right before the collision and 2.0 m/s to the left afterwards. If the second astronaut moves 5.0 m/s to the right after the perfectly elastic collision, what was the second astronaut's initial velocity?

4. Speeds as high as 273 km/h have been recorded for golf balls. Suppose a golf ball whose mass is 45.0 g is moving to the right at 273 km/h and strikes another ball that is at rest. If after the perfectly elastic collision the golf ball moves 91 km/h to the left and the other ball moves 182 km/h to the right, what is the mass of the second ball?

5. Jana Novotna of what is now the Czech Republic has the strongest serve among her fellow tennis players. In 1993, she sent the ball flying with a speed of 185 km/h. Suppose a tennis ball moving to the right at this speed hits a moveable target of unknown mass. After the one-dimensional, perfectly elastic collision, the tennis ball bounces to the left with a speed of 80.0 km/h. If the tennis ball's mass is 5.70×10^{-2} kg, what is the target's mass? (Hint: Use the conservation of kinetic energy to solve for the second unknown quantity.)

6. Recall the two colliding snow trains in item 5 of the previous section. Suppose now that the collision between the two trains is perfectly elastic instead of inelastic. The train with a 4.00×10^5 kg and a velocity of 32.0 km/h to the right is struck from behind by a second train with a mass of 1.60×10^5 kg and a velocity of 36.0 km/h to the right. If the first train's velocity increases to 35.5 km/h to the right, what is the final velocity of the second train after the collision?

7. A dump truck used in Canada has a mass of 5.50×10^5 kg when loaded and 2.30×10^5 kg when empty. Suppose two such trucks, one loaded and one empty, crash into each other at a monster truck show. The trucks are supplied with special bumpers that make a collision almost perfectly elastic. If the trucks hit each other at equal speeds of 5.00 m/s and the less massive truck recoils to the right with a speed of 9.10 m/s, what is the velocity of the full truck after the collision?

NAME ______________________ DATE ____________ CLASS ____________

Circular Motion and Gravitation

Problem A

CENTRIPETAL ACCELERATION

PROBLEM

Calculate the orbital radius of the Earth, if its tangential speed is 29.7 km/s and the centripetal acceleration acting on Earth is 5.9×10^{-3} m/s^2.

SOLUTION

Given: $v_t = 29.7$ km/s $\qquad a_c = 5.9 \times 10^{-3}$ m/s^2

Unknown: $r = ?$

Use the centripetal acceleration equation written in terms of tangential speed. Rearrange the equation to solve for r.

$$a_c = \frac{v_t^2}{r}$$

$$r = \frac{v_t^2}{a_c} = \frac{(29.7 \times 10^3 \text{ m/s})^2}{5.9 \times 10^{-3} \text{ m/s}^2} = \boxed{1.5 \times 10^{11} \text{ m} = 1.5 \times 10^8 \text{ km}}$$

ADDITIONAL PRACTICE

1. The largest salami in the world, made in Norway, was more than 20 m long. If a hungry mouse ran around the salami's circumference with a tangential speed of 0.17 m/s, the centripetal acceleration of the mouse was 0.29 m/s^2. What was the radius of the salami?
2. An astronomer at the equator measures the Doppler shift of sunlight at sunset. From this, she calculates that Earth's tangential velocity at the equator is 465 m/s. The centripetal acceleration at the equator is 3.41×10^{-2} m/s^2. Use this data to calculate Earth's radius.
3. In 1994, Susan Williams of California blew a bubble-gum bubble with a diameter of 58.4 cm. If this bubble were rigid and the centripetal acceleration of the equatorial points of the bubble were 8.50×10^{-2} m/s^2, what would the tangential speed of those points be?
4. An ostrich lays the largest bird egg. A typical diameter for an ostrich egg at its widest part is 12 cm. Suppose an egg of this size rolls down a slope so that the centripetal acceleration of the shell at its widest part is 0.28 m/s^2. What is the tangential speed of that part of the shell?
5. A waterwheel built in Hamah, Syria, has a radius of 20.0 m. If the tangential velocity at the wheel's edge is 7.85 m/s, what is the centripetal acceleration of the wheel?
6. In 1995, Cathy Marsal of France cycled 47.112 km in 1.000 hour. Calculate the magnitude of the centripetal acceleration of Marsal with respect to Earth's center. Neglect Earth's rotation, and use 6.37×10^3 km as Earth's radius.

NAME ______________________ DATE ______________ CLASS ______________

Circular Motion and Gravitation

Problem B

CENTRIPETAL FORCE

PROBLEM

The royal antelope of western Africa has an average mass of only 3.2 kg. Suppose this antelope runs in a circle with a radius of 30.0 m. If a force of 8.8 N maintains this circular motion, what is the antelope's tangential speed?

SOLUTION

Given: $m = 3.2$ kg

$r = 30.0$ m

$F_c = 8.8$ N

Unknown: $v_t = ?$

Use the equation for centripetal force, and rearrange it to solve for tangential speed.

$$F_c = \frac{mv_t^2}{r}$$

$$v_t = \sqrt{\frac{F_c r}{m}} = \sqrt{\frac{(8.8\ \text{N})(30.0\ \text{m})}{3.2\ \text{kg}}} = \sqrt{82\ \frac{\text{m}^2}{\text{s}^2}}$$

$$v_t = \boxed{9.1\ \text{m/s}}$$

ADDITIONAL PRACTICE

1. Gregg Reid of Atlanta, Georgia, built a motorcycle that is over 4.5 m long and has a mass of 235 kg. The force that holds Reid and his motorcycle in a circular path with a radius 25.0 m is 1850 N. What is Reid's tangential speed? Assume Reid's mass is 72 kg.
2. With an average mass of only 30.0 g, the mouse lemur of Madagascar is the smallest primate on Earth. Suppose this lemur swings on a light vine with a length of 2.4 m, so that the tension in the vine at the bottom point of the swing is 0.393 N. What is the lemur's tangential speed at that point?
3. In 1994, Mata Jagdamba of India had very long hair. It was 4.23 m long. Suppose Mata conducted experiments with her hair. First, she determined that one hair strand could support a mass of 25 g. She then attached a smaller mass to the same hair strand and swung it in the horizontal plane. If the strand broke when the tangential speed of the mass reached 8.1 m/s, how large was the mass?
4. Pat Kinch used a racing cycle to travel 75.57 km/h. Suppose Kinch moved at this speed around a circular track. If the combined mass of Kinch and the cycle was 92.0 kg and the average centripetal force was 12.8 N, what was the radius of the track?

5. In 1992, a team of 12 athletes from Great Britain and Canada rappelled 446 m down the CN Tower in Toronto, Canada. Suppose an athlete with a mass of 75.0 kg, having reached the ground, took a joyful swing on the 446 m-long rope. If the speed of the athlete at the bottom point of the swing was 12 m/s, what was the centripetal force? What was the tension in the rope? Neglect the rope's mass.

Circular Motion and Gravitation

Problem C

GRAVITATIONAL FORCE

PROBLEM

The sun has a mass of 2.0×10^{30} kg and a radius of 7.0×10^5 km. What mass must be located at the sun's surface for a gravitational force of 470 N to exist between the mass and the sun?

SOLUTION

Given:

$m_1 = 2.0 \times 10^{30}$ kg

$r = 7.0 \times 10^5$ km $= 7.0 \times 10^8$ m

$G = 6.673 \times 10^{-11}$ N•m²/kg²

$F_g = 470$ N

Unknown: $m_2 = ?$

Use Newton's universal law of gravitation, and rearrange it to solve for the second mass.

$$F_g = G\frac{m_1 m_2}{r^2}$$

$$m_2 = \frac{F_g r^2}{G m_1} = \frac{(470\ \text{N})(7.0 \times 10^8\ \text{m})^2}{\left(6.673 \times 10^{-11}\ \frac{\text{N•m}^2}{\text{kg}^2}\right)(2.0 \times 10^{30}\ \text{kg})}$$

$$m_2 = \boxed{1.7\ \text{kg}}$$

ADDITIONAL PRACTICE

1. Deimos, a satellite of Mars, has an average radius of 6.3 km. If the gravitational force between Deimos and a 3.0 kg rock at its surface is 2.5×10^{-2} N, what is the mass of Deimos?

2. A 3.08×10^4 kg meteorite is on exhibit in New York City. Suppose this meteorite and another meteorite are separated by 1.27×10^7 m (a distance equal to Earth's average diameter). If the gravitational force between them is 2.88×10^{-16} N, what is the mass of the second meteorite?

3. In 1989, a cake with a mass of 5.81×10^4 kg was baked in Alabama. Suppose a cook stood 25.0 m from the cake. The gravitational force exerted between the cook and the cake was 5.0×10^{-7} N. What was the cook's mass?

4. The largest diamond ever found has a mass of 621 g. If the force of gravitational attraction between this diamond and a person with a mass of 65.0 kg is 1.0×10^{-12} N, what is the distance between them?

5. The passenger liners *Carnival Destiny* and *Grand Princess*, built recently, have a mass of about 1.0×10^8 kg each. How far apart must these two ships be to exert a gravitational attraction of 1.0×10^{-3} N on each other?

6. In 1874, a swarm of locusts descended on Nebraska. The swarm's mass was estimated to be 25×10^9 kg. If this swarm were split in half and the halves separated by 1.0×10^3 km, what would the magnitude of the gravitational force between the halves be?

7. Jupiter, the largest planet in the solar system, has a mass 318 times that of Earth and a volume that is 1323 times greater than Earth's. Calculate the magnitude of the gravitational force exerted on a 50.0 kg mass on Jupiter's surface.

Circular Motion and Gravitation

Problem D

PERIOD AND SPEED OF AN ORBITING OBJECT

PROBLEM

A satellite in geostationary orbit rotates at exactly the same rate as Earth, so the satellite always remains in the same position relative to Earth's surface. The period of Earth's rotation is 23 hours, 56 minutes, and 4 seconds. What is the altitude of a satellite in geostationary orbit?

SOLUTION

1. DEFINE **Given:** $T = 23{:}56{:}04 = 86\ 164$ s

Unknown: $r_s = ?$

2. PLAN **Choose an equation or situation:** Use the equation for the period of an object in circular orbit, and rearrange the equation to solve for r.

$$T = 2\pi\sqrt{\frac{r^3}{Gm}}$$

$$T^2 = 4\pi^2\frac{r^3}{Gm}$$

$$r^3 = \frac{GmT^2}{4\pi^2}$$

$$r = \sqrt[3]{\frac{GmT^2}{4\pi^2}}$$

Use **Table 1** in your textbook to find Earth's mass (m) and radius (r_e).

$r_e = 6.38 \times 10^6$ m (from Table 1)

$m = 5.97 \times 10^{24}$ kg (from Table 1)

Once you have found the total radius of the orbit (r), you can subtract Earth's radius (r_e) to find the altitude of the satellite (r_s).

$$r_s = r - r_e$$

3. CALCULATE **Substitute the values into the equations and solve:**

$$r = \sqrt[3]{\frac{\left(6.673 \times 10^{-11}\ \text{N}\bullet\frac{\text{m}^2}{\text{kg}^2}\right)(5.97 \times 10^{24}\ \text{kg})\ (86\ 164\ \text{s})^2}{4\pi^2}} = 4.22 \times 10^7\text{m}$$

$$r_s = r - r_e = (4.22 \times 10^7\ \text{m}) - (6.38 \times 10^6\ \text{m}) = \boxed{3.58 \times 10^7\ \text{m}}$$

4. EVALUATE A satellite in geostationary orbit is at an altitude of 3.58×10^7 m = 35,800 km.

NAME ______________________ DATE ______________ CLASS ______________

ADDITIONAL PRACTICE

1. The period of Mars' rotation is 24 hours, 37 minutes, and 23 seconds. At what altitude above Mars would a "Mars-stationary" satellite orbit?
2. Pluto's moon, Charon, has an orbital period of 153 hours. How far is Charon from Pluto?
3. The orbital radius of a satellite in geostationary orbit is 4.22×10^7 m (see sample problem on previous page). What is the orbital speed of a satellite in geostationary orbit?
4. Earth's moon orbits Earth at a mean distance of 3.84×10^8 m. What is the moon's orbital speed?
5. Earth's moon orbits Earth at a mean distance of 3.84×10^8 m. What is the moon's orbital period? Express your answer in Earth days.
6. Use data from **Table 1** in your textbook to calculate the length of Neptune's "year" (the period of its orbit around the Sun). Express your answer in Earth years.
7. The asteroid (45)Eugenia has a small moon named S/1998(45)1. The moon orbits Eugenia once every 4.7 days at a distance of 1.19×10^3 km. What is the mass of (45)Eugenia?
8. V404 Cygni is a dark object orbited by a star in the constellation Cygnus. Many astronomers believe the object is a black hole. Suppose the star's orbit has a mean radius of 2.30×10^{10} m and a period of 6.47 days. What is the mass of the black hole? How many times larger is the mass of the black hole than the mass of the sun?

NAME ______________________ DATE ____________ CLASS ____________

Circular Motion and Gravitation

Problem E

TORQUE

PROBLEM

A beam that is hinged near one end can be lowered to stop traffic at a railroad crossing or border checkpoint. Consider a beam with a mass of 12.0 kg that is partially balanced by a 20.0 kg counterweight. The counterweight is located 0.750 m from the beam's fulcrum. A downward force of 1.60×10^2 N applied over the counterweight causes the beam to move upward. If the net torque on the beam is 29.0 N·m when the beam makes an angle of 25.0° with respect to the ground, how long is the beam's longer section? Assume that the portion of the beam between the counterweight and fulcrum has no mass.

SOLUTION

1. DEFINE **Given:**

$m_b = 12.0$ kg

$m_c = 20.0$ kg

$d_c = 0.750$ m

$F_{applied} = 1.60 \times 10^2$ N

$\tau_{net} = 29.0$ N•m

$\theta = 90.0° - 25.0° = 65.0°$

$g = 9.81$ m/s^2

Unknown: $\ell = ?$

Diagram:

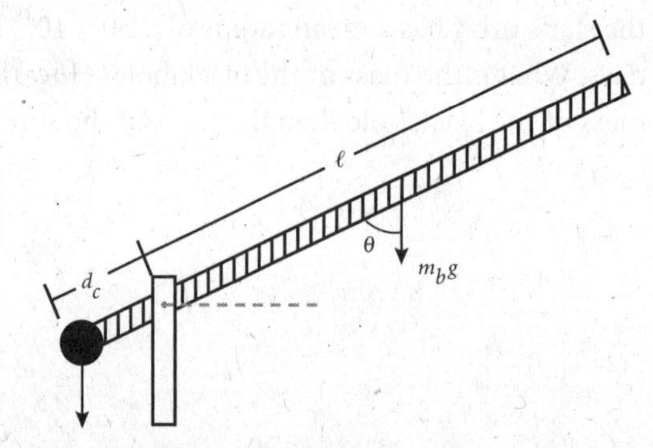

2. PLAN **Choose the equation(s) or situation:** Apply the definition of torque to each force and add up the individual torques.

$$\tau = F\,d\,(\sin\theta)$$

$$\tau_{net} = \tau_a + \tau_b + \tau_c$$

where τ_a = counterclockwise torque produced by applied force = $F_{applied}\,d_c\,(\sin\theta)$

τ_b = clockwise torque produced by weight of beam

$= -m_b\,g\left(\frac{\ell}{2}\right)(\sin\theta)$

τ_c = counterclockwise torque produced by counterweight

$= m_c\,g\,d_c\,(\sin\theta)$

$$\tau_{net} = F_{applied}\,d_c\,(\sin\theta) - m_b\,g\left(\frac{\ell}{2}\right)(\sin\theta) + m_c\,g\,d_c\,(\sin\theta)$$

Note that the clockwise torque is negative, while the counterclockwise torques are positive.

Rearrange the equation(s) to isolate the unknown(s):

$$m_b g\left(\frac{\ell}{2}\right) = (F_{applied} + m_c g)\, d_c - \left(\frac{\tau_{net}}{\sin\theta}\right)$$

$$\ell = \frac{2\left[(F_{applied} + m_c g)d_c - \left(\frac{\tau_{net}}{\sin\theta}\right)\right]}{m_b g}$$

3. CALCULATE **Substitute the values into the equation(s) and solve:**

$$\ell = \frac{(2)([1.60 \times 10^2 \text{ N} + (20.0 \text{ kg})(9.81 \text{ m/s}^2)]\,(0.750 \text{ m}) - \left[\frac{29.0 \text{ N}\bullet\text{m}}{\sin 65.0°}\right]}{(12.0 \text{ kg})(9.81 \text{ m/s}^2)}$$

$$\ell = \frac{(2)\,[(1.60 \times 10^2 \text{ N} + 196 \text{ N})(0.750 \text{ m}) - 32.0 \text{ N}\bullet\text{m}]}{(12.0 \text{ kg})(9.81 \text{ m/s}^2)}$$

$$\ell = \frac{(2)\,[356 \text{ N})(0.750 \text{ m}) - 32.0 \text{ N}\bullet\text{m}]}{(12.0 \text{ kg})(9.81 \text{ m/s}^2)}$$

$$\ell = \frac{(2)(2.67 \times 10^2 \text{ N}\bullet\text{m} - 32.0 \text{ N}\bullet\text{m}]}{(12.00 \text{ kg})(9.81 \text{ m/s}^2)}$$

$$\ell = \frac{(2)(235 \text{ N}\bullet\text{m})}{(12.0 \text{ kg})(9.81 \text{ m/s}^2)}$$

$$\ell = \boxed{3.99 \text{ m}}$$

4. EVALUATE For a constant applied force, the net torque is greatest when θ is 90.0° and decreases as the beam rises. Therefore, the beam rises fastest initially.

ADDITIONAL PRACTICE

1. The nests built by the mallee fowl of Australia can have masses as large as 3.00×10^5 kg. Suppose a nest with this mass is being lifted by a crane. The boom of the crane makes an angle of 45.0° with the ground. If the axis of rotation is the lower end of the boom at point *A*, the torque produced by the nest has a magnitude of 3.20×10^7 N•m. Treat the boom's mass as negligible, and calculate the length of the boom.
2. The pterosaur was the most massive flying dinosaur. The average mass for a pterosaur has been estimated from skeletons to have been between 80.0 and 120.0 kg. The wingspan of a pterosaur was greater than 10.0 m. Suppose two pterosaurs with masses of 80.0 kg and 120.0 kg sat on the middle and the far end, respectively, of a light horizontal tree branch. The pterosaurs produced a net counterclockwise torque of 9.4 kN•m about the end of the branch that was attached to the tree. What was the length of the branch?

3. A meterstick of negligible mass is fixed horizontally at its 100.0 cm mark. Imagine this meterstick used as a display for some fruits and vegetables with record-breaking masses. A lemon with a mass of 3.9 kg hangs from the 70.0 cm mark, and a cucumber with a mass of 9.1 kg hangs from the x cm mark. What is the value of x if the net torque acting on the meterstick is 56.0 N•m in the counterclockwise direction?

4. In 1943, there was a gorilla named N'gagi at the San Diego Zoo. Suppose N'gagi were to hang from a bar. If N'gagi produced a torque of -1.3×10^4 N•m about point *A*, what was his weight? Assume the bar has negligible mass.

5. The first—and, in terms of the number of passengers it could carry, the largest—Ferris wheel ever constructed had a diameter of 76 m and held 36 cars, each carrying 60 passengers. Suppose the magnitude of the torque, produced by a ferris wheel car and acting about the center of the wheel, is -1.45×10^6 N•m. What is the car's weight?

6. In 1897, a pair of huge elephant tusks were obtained in Kenya. One tusk had a mass of 102 kg, and the other tusk's mass was 109 kg. Suppose both tusks hang from a light horizontal bar with a length of 3.00 m. The first tusk is placed 0.80 m away from the end of the bar, and the second, more massive tusk is placed 1.80 m away from the end. What is the net torque produced by the tusks if the axis of rotation is at the center of the bar? Neglect the bar's mass.

7. A catapult, a device used to hurl heavy objects such as large stones, consists of a long wooden beam that is mounted so that one end of it pivots freely in a vertical arc. The other end of the beam consists of a large hollowed bowl in which projectiles are placed. Suppose a catapult provides an angular acceleration of 50.0 rad/s^2 to a 5.00×10^2 kg boulder. This can be achieved if the net torque acting on the catapult beam, which is 5.00 m long, is 6.25×10^5 N•m.

a. If the catapult is pulled back so that the beam makes an angle of 10.0° with the horizontal, what is the magnitude of the torque produced by the 5.00×10^2 kg boulder?

b. If the force that accelerates the beam and boulder acts perpendicularly on the beam 4.00 m from the pivot, how large must that force be to produce a net torque of 6.25×10^5 N•m?

NAME ______________________ DATE ____________ CLASS ______________

Fluid Mechanics

Problem A

BUOYANT FORCE

PROBLEM

The highest natural concentration of salts in water are found in the evaporating remnants of old oceans, such as the Dead Sea in Israel. Suppose a swimmer with a volume of 0.75 m^3 is able to float just beneath the surface of water with a density of 1.02×10^3 kg/m^3. How much extra mass can the swimmer carry and be able to float just beneath the surface of the Dead Sea, which has a density of 1.22×10^3 kg/m^3?

SOLUTION

1. DEFINE **Given:**

$V = 0.75$ m^3

$\rho_1 = 1.02 \times 10^3$ kg/m^3

$\rho_2 = 1.22 \times 10^3$ kg/m^3

Unknown: $m' = ?$

2. PLAN **Choose the equation(s) or situation:** In both bodies of water, the buoyant force equals the weight of the floating object.

$$F_{B,1} = F_{g,1}$$
$$\rho_1 V_g = m_g$$
$$F_{B,2} = F_{g,2}$$
$$\rho_2 V_g = (m + m')g = \rho_1 V_g + m'g$$

Rearrange the equation(s) to isolate the unknown(s):

$$m' = (\rho_2 - \rho_1)V$$

3. CALCULATE **Substitute the values into the equation(s) and solve:**

$$m' = (1.22 \times 10^3 \text{ kg/m}^3 - 1.02 \times 10^3 \text{ kg/m}^3)(0.75 \text{ m}^3)$$
$$m' = (0.20 \times 10^3 \text{ kg/m}^3)(0.75 \text{ m}^3)$$
$$m' = \boxed{150 \text{ kg}}$$

4. EVALUATE The mass that can be supported by buoyant force increases with the difference in fluid densities.

ADDITIONAL PRACTICE

1. The heaviest pig ever raised had a mass of 1158 kg. Suppose you placed this pig on a raft made of dry wood. The raft completely submerged in water so that the raft's top surface was just level with the surface of the lake. If the raft's volume was 3.40 m^3, what was the mass of the raft's dry wood? The density of fresh water is 1.00×10^3 kg/m^3.

2. *La Belle*, one of four ships that Robert La Salle used to establish a French colony late in the seventeenth century, sank off the coast of Texas. The ship's well-preserved remains were discovered and excavated in the 1990s. Among those remains was a small bronze cannon, called a minion. Suppose the minion's total volume is 4.14×10^{-2} m^3. What is the minion's mass if its apparent weight in sea water is 3.115×10^3 N? The density of sea water is 1.025×10^3 kg/m^3.

3. William Smith built a small submarine capable of diving as deep as 30.0 m. The submarine's volume can be approximated by that of a cylinder with a length of 3.00 m and a cross-sectional area of 0.500 m^2. Suppose this submarine dives in a freshwater river and then moves out to sea, which naturally consists of salt water. What mass of fresh water must be added to the ballast to keep the submarine submerged? The density of fresh water is 1.000×10^3 kg/m^3, and the density of sea water is 1.025×10^3 kg/m^3.

4. The largest iceberg ever observed had an area of 3.10×10^4 km^2, which is larger than the area of Belgium. If the top and bottom surfaces of the iceberg were flat and the thickness of the submerged part was 0.84 km, how large was the buoyant force acting on the iceberg? The density of sea water equals 1.025×10^3 kg/m^3.

5. A cannon built in 1868 in Russia could fire a cannonball with a mass of 4.80×10^2 kg and a radius of 0.250 m. When suspended from a scale and submerged in water, a cannonball of this type has an apparent weight of 4.07×10^3 N. How large is the buoyant force acting on the cannonball? The density of fresh water is 1.00×10^3 kg/m^3.

6. The tallest iceberg ever measured stood 167 m above the water. Suppose that both the top and the bottom of this iceberg were flat and the thickness of the submerged part was estimated to be 1.50 km. Calculate the density of the ice. The density of sea water equals 1.025×10^3 kg/m^3.

7. The Russian submarines of the "Typhoon" class are the largest submarines in the world. They have a length of 1.70×10^2 m and an average diameter of 13.9 m. When submerged, they displace 2.65×10^7 kg of sea water. Assume that these submarines have a simple cylindrical shape. If a "Typhoon"-class submarine has taken on enough ballast that it descends with a net acceleration of 2.00 m/s^2, what is the submarine's density during its descent?

8. To keep Robert La Salle's ship *La Belle* well preserved, shipbuilders are reconstructing the ship in a large tank filled with fresh water. Polyethylene glycol, or PEG, will then be slowly added to the water until a 30 percent PEG solution is formed. Suppose *La Belle* displaces 6.00 m^3 of liquid when submerged. If the ship's apparent weight decreases by 800 N as the PEG concentration increases from 0 to 30 percent, what is the density of the final PEG solution? The density of fresh water is 1.00×10^3 kg/m^3.

NAME ______________________ DATE ______________ CLASS ______________

Fluid Mechanics

Problem B

PRESSURE

PROBLEM

The largest helicopter in the world, which was built in Russia, has a mass of 1.03×10^3 kg. If you placed this helicopter on a large piston of a hydraulic lift, what force would need to be applied to the small piston in order to slowly lift the helicopter? Assume that the weight of the helicopter is distributed evenly over the large piston's area, which is 1.40×10^2 m^2. The area of the small piston is 0.80 m^2.

SOLUTION

Given: $m = 1.03 \times 10^5$ kg

$A_1 = 1.40 \times 10^2$ m^2

$A_2 = 0.80$ m^2

$g = 9.81$ m/s^2

Unknown: $F_2 = ?$

Use the equation for pressure to equate the two opposing pressures in terms of force and area.

$$P = \frac{F}{A} \qquad P_1 = P_2$$

$$\frac{F_1}{A_1} = \frac{F_2}{A_2}$$

$$F_2 = F_1\frac{A_2}{A_1} = mg\left(\frac{A_2}{A_1}\right)$$

$$F_2 = (1.03 \times 10^5 \text{ kg})(9.81 \text{ m/s}^2)\left(\frac{0.80 \text{ m}^2}{1.40 \times 10^2 \text{ m}^2}\right)$$

$$F_2 = \boxed{5.8 \times 10^3 \text{ N}}$$

ADDITIONAL PRACTICE

1. Astronauts and cosmonauts have used pressurized spacesuits to explore the low-pressure regions of space. The pressure inside one of these suits must be close to that of Earth's atmosphere at sea level so that the space explorer may be safe and comfortable. The pressure on the outside of the suit is a fraction of 1.0 Pa. Clearly, pressurized suits are made of extremely sturdy material that can tolerate the stress from these pressure differences. If the average interior surface area of a pressurized spacesuit is 3.3 m^2, what is the force exerted on the suit's material? Assume that the pressure outside the suit is zero and that the pressure inside the suit is 1.01×10^5 Pa.

2. A strange idea to control volcanic eruptions is developed by a daydreaming engineer. The engineer imagines a giant piston that fits into the volcano's shaft, which leads from Earth's surface down to the magma chamber. The piston controls an eruption by exerting pressure that is equal to or greater than the pressure of the hot gases, ash, and magma that rise from the magma chamber through the shaft. The engineer assumes that the pressure of the volcanic material is 4.0×10^{11} Pa, which is the pressure in Earth's interior. If the material rises into a cylindrical shaft with a radius of 50.0 m, what force is needed on the other side of the piston to balance the pressure of the volcanic material?

3. The largest goat ever grown on a farm had a mass of 181 kg; on the other hand, the smallest "pygmy" goats have a mass of only about 16 kg. Imagine an agricultural show in which a large goat with a mass of 181 kg exerts a pressure on a hydraulic-lift piston that is equal to the pressure exerted by three pygmy goats, each of which has a mass of 16.0 kg. The area of the piston on which the large goat stands is 1.8 m^2. What is the area of the piston on which the pygmy goats stand?

4. The greatest load ever raised was the offshore Ekofisk complex in the North Sea. The complex, which had a mass of 4.0×10^7 kg, was raised 6.5 m by more than 100 hydraulic jacks. Imagine that his load could have been raised using a single huge hydraulic lift. If the load had been placed on the large piston and a force of 1.2×10^4 N had been applied to the small piston, which had an area of 5.0 m^2, what must the large piston's area have been?

5. The pressure that can exist in the interior of a star due to the weight of the outer layers of hot gas is typically several hundred billion times greater than the pressure exerted on Earth's surface by Earth's atmosphere. Suppose a pressure equal to that estimated for the sun's interior (2.0×10^{16} Pa) acts on a spherical surface within a star. If a force of 1.02 N $\times 10^{31}$ N produces this pressure, what is the area of the surface? What is the sphere's radius r? (Recall that a sphere's surface area equals $4\pi r^2$.)

6. The eye of a giant squid can be more than 35 cm in diameter—the largest eye of any animal. Giant squid also live at depths greater than a mile below the ocean's surface. At a depth of 2 km, the outer half of a giant squid's eye is acted on by an external force of 4.6×10^6 N. Assuming the squid's eye has a diameter of 38 cm, what is the pressure on the eye? (Hint: Treat the eye as a sphere.)

7. The largest tires in the world, which are used for certain huge dump trucks, have diameters of about 3.50 m. Suppose one of these tires has a volume of 5.25 m^3 and a surface area of 26.3 m^2. If a force of 1.58×10^7 N acts on the inner area of the tire, what is the absolute pressure inside the tire? What is the gauge pressure on the tire's surface?

NAME ______________ DATE ______________ CLASS ______________

Fluid Mechanics

Problem C

PRESSURE AS A FUNCTION OF DEPTH

PROBLEM

In 1969, a whale dove and remained underwater for nearly 2 h. Evidence indicates that the whale reached a depth of 3.00 km, where the whale sustained a pressure of 3.03×10^7 Pa. Estimate the density of sea water.

SOLUTION

Given: $h = 3.00 \text{ km} = 3.00 \times 10^3 \text{ m}$

$P = 3.03 \times 10^7 \text{ Pa}$

$P_o = 1.01 \times 10^5 \text{ Pa}$

$g = 9.81 \text{ m/s}^2$

Unknown: $\rho = ?$

Use the equation for fluid pressure as a function of depth, and rearrange it to solve for density.

$$P = P_o + \rho g h$$

$$\rho = \frac{P - P_o}{gh}$$

$$\rho = \frac{(3.03 \times 10^7 \text{ Pa}) - (1.01 \times 10^5 \text{ Pa})}{(9.81 \text{ m/s}^2)(3.00 \times 10^3 \text{ m})}$$

$$\rho = \frac{3.02 \times 10^7 \text{ Pa}}{(9.81 \text{ m/s}^2)(3.00 \times 10^3 \text{ m})}$$

$$\rho = \boxed{1.03 \times 10^3 \text{ kg/m}^3}$$

ADDITIONAL PRACTICE

1. In 1994, a 16.8 m tall oil-filled barometer was constructed in Belgium. Suppose the barometer column was 80.0 percent filled with oil. What was the density of the oil if the pressure at the bottom of the column was 2.22×10^5 Pa and the air pressure at the top of the oil column was 1.01×10^5 Pa?

2. One of the lowest atmospheric pressures ever measured at sea level was 8.88×10^4 Pa, which existed in hurricane Gilbert in 1988. This same pressure can be found at a height 950 m above sea level. Use this information to estimate the density of air.

3. In 1993, Francisco Ferreras of Cuba held his breath and took a dive that lasted more than 2 min. The maximum pressure Ferreras experienced was 13.6 times greater than atmospheric pressure. To what depth did Ferreras dive? The density of sea water is $1.025 \times 10^3 \text{ kg/m}^3$.

4. A penguin can endure pressures as great as 4.90×10^6 Pa. What is the maximum depth to which a penguin can dive in sea water?

5. In 1942, the British ship *Edinburgh,* which was carrying a load of 460 gold ingots, sank off the coast of Norway. In 1981, all of the gold was recovered from a depth of 245 m by a team of 12 divers. What was the pressure exerted by the ocean's water at that depth?

6. In 1960, a bathyscaph descended 10 916 m below the ocean's surface. What was the pressure exerted on the bathyscaph at that depth?

NAME ______________________ DATE ____________ CLASS ____________

Heat

Problem A

TEMPERATURE CONVERSION

PROBLEM

The temperature at the surface of the sun is estimated to be 5.97×10^3 K. Express this temperature in degrees Fahrenheit and in degrees Celsius.

SOLUTION

Given: $T = 5.97 \times 10^3 \text{K}$

Unknown: $T_F = ?$ $T_C = ?$

Use the Celsius-Fahrenheit and Celsius-Kelvin temperature conversion equations.

$$T_C = T - 273.15 = (5.97 \times 10^3 - 0.27 \times 10^3)\text{K} = \boxed{5.70 \times 10^3 \text{ K}}$$

$$T_F = \tfrac{9}{5}T_C + 32.0 = \left[\tfrac{9}{5}(5.70 \times 10^3) + 32.0\right]{}^\circ\text{F} = \boxed{1.03 \times 10^4 {}^\circ\text{F}}$$

ADDITIONAL PRACTICE

1. Usually, people die if their body temperature drops below 35°C. There was one case, however, of a two-year-old girl who had been accidentally locked outside in the winter. She survived, even though her body temperature dropped as low as 14°C. Express this temperature in kelvins and in degrees Fahrenheit.

2. In experiments conducted by the United States Air Force, subjects endured air temperatures of 4.00×10^2°F. Express this temperature in degrees Celsius and in kelvins.

3. The temperature of the moon's surface can reach 117°C when exposed to the sun and can cool to −163°C when facing away from the sun. Express this temperature change in degrees Fahrenheit.

4. Because of Venus' proximity to the sun and its thick, high-pressure atmosphere, its temperature can rise to 860.0°F. Express this temperature in degrees Celsius.

5. On January 22, 1943, the air temperature at Spearfish, South Dakota, rose 49.0°F in 2 min to reach a high temperature of 7.00°C. What were the initial and final temperatures in degrees Fahrenheit? What was the temperature in degrees Celsius before the temperature increase?

6. In 1916, Browning, Montana, experienced a temperature decrease of 56°C during a 24 h period. The final temperature was −49°C. Express in kelvins the temperatures at the beginning and the end of the 24 h period.

7. In 1980, Willie Jones of Atlanta, Georgia, was hospitalized with heatstroke, having a body temperature of 116°F. Fortunately, he survived. Express Willie's body temperature in kelvins.

Heat

Problem B

CONSERVATION OF ENERGY

PROBLEM

The moon has a mass of 7.3×10^{22} kg and an average orbital speed of 1.02×10^3 m/s. Suppose all of the moon's kinetic energy goes into increasing the internal energy of a quantity of water at a temperature of 100.0°C. If it takes 2.26×10^6 J to vaporize 1.00 kg of water that is initially at 100.0°C, what mass of this water can be vaporized?

SOLUTION

1. DEFINE **Given:**

m_m = mass of moon = 7.3×10^{22} kg

v_m = speed of moon = 1.02×10^3 m/s

k = energy needed to vaporize 1 kg of water = 2.26×10^6 J

Unknown: m_w = mass of water vaporized = ?

2. PLAN **Choose the equation(s) or situation:** Use the equation for energy conservation, including a term for the change in internal energy.

$$\Delta PE + \Delta KE + \Delta U = 0$$

The moon's kinetic energy alone is taken into consideration. Therefore ΔPE equals zero. Because all of the moon's kinetic energy is transferred to the water's internal energy, the moon's final kinetic energy is also zero.

$$\Delta KE + \Delta U = KE_f - KE_i + \Delta U = 0 - KE_i + \Delta U = 0$$

$$\Delta U = KE_i = \tfrac{1}{2}m_m(v_m)^2$$

To find the mass of water that will be vaporized, the change in internal energy must be divided by the conversion constant, k.

$$m_w = \frac{\Delta U}{k} = \frac{m_m(v_m)^2}{2k}$$

3. CALCULATE **Substitute the values into the equation(s) and solve:**

$$m_w = \frac{(7.3 \times 10^{22}\ \text{kg})(1.02 \times 10^3\ \text{m/s})^2}{2(2.26 \times 10^6\ \text{J/kg})} = \boxed{1.7 \times 10^{22}\ \text{kg}}$$

4. EVALUATE The mass of the water vaporized is only 23 percent of the moon's mass. This suggests that a smaller object moving with the moon's speed would vaporize a mass of water only 0.23 times as large as its own mass. Because the water is already at its boiling point, this result indicates that a considerable amount of energy goes into the process of vaporization.

ADDITIONAL PRACTICE

1. A British-built Hovercraft™—a vehicle that cruises on a cushion of air—has a mass of 3.05×10^5 kg and can attain a speed of 120.0 km/h. Suppose this vehicle slows down from 120.0 km/h to 90.0 km/h and that the

change in its kinetic energy is used to raise the temperature of a quantity of water by 10.0°C. Knowing that 4186 J is required to raise the temperature of 1.00 kg of water by 1.00°C, calculate the mass of the heated water.

2. The Westin Stamford Hotel in Detroit is 228 m tall. Suppose a piece of ice, which initially has a temperature of 0.0°C, falls from the hotel roof and crashes to the ground. Assuming that 50.0 percent of the ice's mechanical energy during the fall and collision is absorbed by the ice and that 3.33×10^5 J is required to melt 1.00 kg of ice, calculate the fraction of the ice's mass that would melt.

3. The French-built *Concorde,* the fastest passenger jet plane, is known to travel with a speed as great as 2.333×10^3 km/h. Suppose the plane travels horizontally at an altitude of 4.000×10^3 m and at maximum speed when a fragment of metal breaks free from the plane. The metal has, of course, the same horizontal speed as the plane, and when it lands on the ground, it will have absorbed 1.00 percent of its total mechanical energy. If it takes 355 J to raise the temperature of 1.00 kg of this metal by 1.00°C, how great a temperature change will the metal fragment experience from the time it breaks free from the *Concorde* to the time it lands on the ground? Ignore air resistance.

4. Mount Everest is the world's highest mountain. Its height is 8848 m. Suppose a steel alpine hook were to slowly slide off the summit of Everest and fall all the way to the base of the mountain. If 20.0 percent of the hook's mechanical energy is absorbed by the hook as internal energy, calculate the final temperature of the hook. Assume that the hook's initial temperature is −18.0°C and that the hook's temperature increases by 1.00°C for every 448 J/kg that is added.

5. The second tallest radio mast in the world is located near Fargo, North Dakota. The tower, which has an overall height of 629 m, was built in just one month by a team of 11 workers. Suppose one of these builders left a 3.00 g copper coin on the top of the tower. During extremely windy weather, the coin falls off the tower and reaches the ground at the tower's base with a speed of 42 m/s. If the coin absorbs 5.0 percent of its total mechanical energy, by how much does the coin's internal energy increase?

6. In 1993, Russell Bradley carried a load of bricks with a total mass of 312 kg up a ramp that had a height of 2.49 m. Suppose Bradley puts the load on the ramp and then pushes the load off the edge with a horizontal speed of 0.50 m/s. If the bricks absorb half their total mechanical energy, how much does their internal energy change?

7. Angel Falls, the highest waterfall in the world, is located in Venezuela. Estimate the height of the waterfall, assuming that the water that falls the complete distance experiences a temperature increase of 0.230°C. In your calculation, assume that the water absorbs 10.0 percent of its mechanical energy and that 4186 J is needed to raise the temperature of 1.00 kg of water by 1.00°C.

Heat

Problem C

CALORIMETRY

PROBLEM

In 1990, the average rate for use of fresh water in the United States was approximately 1.50×10^7 kg each second. Suppose a group of teenagers build a *really* big calorimeter and that they place in it a mass of water equal to the mass of fresh water consumed in 1.00 s. They then use a test sample of gold with a mass equal to that of the United States gold reserve for 1992. Initially, the gold has a temperature of 80.0°C and the water has a temperature of 1.00°C. If the final equilibrium temperature of the gold and water is 2.30°C, what is the mass of the gold?

SOLUTION

1. DEFINE **Given:**

$T_f = 2.30°C$ $\quad m_{water} = m_w = 1.50 \times 10^7$ kg

$T_{gold} = T_g = 80.0°C$ $\quad T_{water} = T_w = 1.00°C$

$c_{p,gold} = c_{p,g} = 129$ J/kg•°C $\quad c_{p,water} = c_{p,w} = 4186$ J/kg•°C

Unknown: $m_{gold} = m_g = ?$

2. PLAN **Choose the equation(s) or situation:** Equate the energy removed from the gold to the energy absorbed by the water.

energy removed from metal = energy absorbed by water

$$-c_{p,g}m_g(T_f - T_g) = c_{p,w}m_w(T_f - T_w)$$

Rearrange the equation(s) to isolate the unknown(s):

$$m_g = \frac{c_{p,w}m_w(T_f - T_w)}{c_{p,g}(T_g - T_f)}$$

3. CALCULATE **Substitute the values into the equation(s) and solve:**

$$m_g = \frac{(4186 \text{ J/kg•°C})(1.50 \times 10^7 \text{ kg})(2.30°C - 1.00°C)}{(129 \text{ J/kg•°C})(80.0°C - 2.30°C)}$$

$$m_g = \frac{(4186 \text{ J/kg•°C})(1.50 \times 10^7 \text{ kg})(1.30°C)}{(129 \text{ J/kg•°C})(77.7°C)} = \boxed{8.14 \times 10^6 \text{ kg}}$$

4. EVALUATE Although the masses of the gold sample and water are unrealistic, the temperatures are entirely reasonable, indicating that temperature is an intrinsic variable of matter that is independent of the quantity of a substance. The small increase in the water's temperature and the large decrease in the gold's temperature is the result of the water having both a larger mass and a larger specific heat capacity.

ADDITIONAL PRACTICE

1. In 1992, the average rate of energy consumption in the United States was about 2.8×10^9 W. Suppose all of the copper produced in the United States in 1992 is placed in the giant calorimeter used in the sample problem. The quantity of energy transferred by heat from the copper to the water is equal to the energy used in the United States during 1.2 s of 1992. If the initial temperature of the copper is 26.0°C, and the final temperature is 21.0°C, what is the mass of the copper?

2. In 1992, a team of firefighters pumped 143×10^3 kg of water in less than four days. What mass of wood can be cooled from a temperature of 280.0°C to one of 100.0°C using this amount of water? Assume that the initial temperature of the water is 20.0°C and that all of the water has a final equilibrium temperature of 100.0°C, but that none of the water is vaporized. Use 1.700×10^3 J/kg•°C for the specific heat capacity of wood.

3. One of the nuclear generators at a power plant in Lithuania has a nominal power of 1.450 GW, making it the most powerful generator in the world. Nippon Steel Corporation in Japan is the world's largest steel producer. Between April 1, 1993, and March, 31, 1994, Nippon Steel's mills produced 25.1×10^9 kg of steel. Suppose this entire quantity of steel is heated and then placed in the giant calorimeter used in the sample problem. If the quantity of energy transferred by heat from the steel to the water equals 1.00 percent of all the energy produced by the Lithuanian generator in a year, what is the temperature change of the steel? Assume that the specific heat capacities of steel and iron are the same.

4. In 1994, to commemorate the 200th anniversary of a beverage company, a giant bottle was constructed and filled with 2.25×10^3 kg of the company's lemonade. Suppose the lemonade has an initial temperature of 28.0°C when 9.00×10^2 kg of ice with a temperature of −18.0°C is added to it. What is the lemonade's temperature at the moment the temperature of the ice reaches 0.0°C? Assume that the lemonade has the same specific heat capacity as water.

5. The water in the Arctic Ocean has a total mass of 1.33×10^{19} kg. The average temperature of the water is estimated to be 4 .000°C. What would the temperature of the water in the Arctic Ocean be if the energy produced in 1.000×10^3 y by the world's largest power plant (1.33×10^{10} W) were transferred by heat to it?

6. There is a little island off the shore of Brazil where the weather is extremely consistent. From 1911 to 1990, the lowest temperature on the island was 18°C (64°F) and the highest temperature was 32°C (90°F). It is known that the liquid in a standard can of soft drink absorbs 20.8 kJ of energy when its temperature increases from 18.0°C to 32.0°C. If the soft drink has a mass of 0.355 kg, what is its specific heat capacity?

7. The lowest temperature ever recorded in Alaska is −62°C. The highest temperature ever recorded in Alaska is 38°C. Suppose a piece of metal with a mass of 180 g and a temperature of −62.0°C is placed in a calorimeter containing 0.500 kg of water with a temperature of 38.0°C. If the final equilibrium temperature of the metal and water is 36.9°C, what is the specific heat capacity of the metal? Use the calculated value of the specific heat capacity and **Table 4** in your textbook to identify the metal.

NAME ______________________ DATE ______________ CLASS ______________

Heat

Problem D

HEAT OF PHASE CHANGE

PROBLEM

The world's deepest gold mine, which is located in South Africa, is over 3 km deep. Every day, the mine transfers enough energy by heat to the mine's cooling systems to melt 3.36×10^7 kg of ice at 0.0°C. If the energy output from the mine is increased by 2.0 percent, to what final temperature will the 3.36×10^7 kg of ice-cold water be heated?

SOLUTION

1. DEFINE

Given:

$m_{ice} = m_{water} = m = 3.36 \times 10^7$ kg

$T_i = 0.0$°C

$L_f = 3.33 \times 10^5$ J/kg

$c_{p,w} = 4186$ J/(kg•°C)

Q' = energy added to water = $(2.0 \times 10^{-2})Q$

Unknown: $T_f = ?$

2. PLAN

Choose the equation(s) or situation: First, determine the amount of energy needed to melt 3.36×10^7 kg of ice by using the equation for the heat of fusion.

$$Q = m_{ice}L_f = mL_f$$

The energy added to the now liquid ice (Q') can then be determined.

$$Q' = (2.0 \times 10^{-2})Q = (2.0 \times 10^{-2})mL_f$$

Finally, the energy added to the water equals the product of the water's mass, specific heat capacity, and change in temperature.

$$Q' = mc_{p,w}(T_f - T_i) = (2.0 \times 10^{-2})mL_f$$

Rearrange the equation(s) to isolate the unknown(s):

$$T_f = \frac{(2.0 \times 10^{-2})mL_f}{mc_{p,w}} + T_i = (2.0 \times 10^{-2})\frac{L_f}{c_{p,w}} + T_i$$

3. CALCULATE

Substitute the values into the equation(s) and solve:

$$T_f = (2.0 \times 10^{-2})\frac{(3.33 \times 10^{-5} \text{ J/kg})}{(4186 \text{ J/kg•°C})} + 0.0\text{°C} = \boxed{1.6\text{°C}}$$

4. EVALUATE

Note that the result is independent of the mass of the ice and water. The amount of energy needed to raise the water's temperature by 1°C is a little more than 1 percent of the energy required to melt the ice.

ADDITIONAL PRACTICE

1. Lake Superior contains about 1.20×10^{16} kg of water, whereas Lake Erie contains only 4.8×10^{14} kg of water. Suppose aliens use these two lakes for cooking. They heat Lake Superior to 100.0°C and freeze Lake Erie to

0.0°C. Then they mix the two lakes together to make a "lake shake." What would be the final temperature of the mixture? Assume that the entire energy transfer by heat occurs between the lakes.

2. The lowest temperature measured on the surface of a planetary body in the solar system is that of Triton, the largest of Neptune's moons. The surface temperature on this distant moon can reach a low of –235°C. Suppose an astronaut brings a water bottle containing 0.500 kg of water to Triton. The water's temperature decreases until the water freezes, then the temperature of the ice decreases until it is in thermal equilibrium with Triton at a temperature of –235°C. If the energy transferred by heat from the water to Triton is 471 kJ, what is the value of the water's initial temperature?

3. Suppose that an ice palace built in Minnesota in 1992 is brought into contact with steam with a temperature of 100.0°C. The temperature and mass of the ice palace are 0.0°C and 4.90×10^6 kg, respectively. If all of the steam liquefies by the time all of the ice melts, what is the mass of the steam?

4. In 1992, 1.804×10^6 kg of silver was produced in the United States. What mass of ice must be melted so that this mass of liquid silver can solidify? Assume that both substances are brought into contact at their melting temperatures. The latent heat of fusion for silver is 8.82×10^4 J/kg.

5. The United States Bullion Depository at Fort Knox, Kentucky, contains almost half a million standard mint gold bars, each with a mass of 12.4414 kg. Assuming an initial bar temperature of 5.0°C, each bar will melt if it absorbs 2.50 MJ of energy transferred by heat. If the specific heat capacity of gold is 129 J/kg•°C and the melting point of gold is 1063°C, calculate the heat of fusion of gold.

6. The world's largest piggy bank has a volume of 7.20 m^3. Suppose the bank is filled with copper pennies and that the pennies occupy 80.0 percent of the bank's total volume. The density of copper is 8.92×10^3 kg/m^3.

a. Find the total mass of the coins in the piggy bank.

b. Consider the mass found in (a). If these copper coins are brought to their melting point, how much energy must be added to the coins in order to melt 15 percent of their mass? The latent heat of fusion for copper is 1.34×10^5 J/kg.

7. The total mass of fresh water on Earth is 3.5×10^{19} kg. Suppose all this water has a temperature of 10.0°C. Suppose the entire energy output of the sun is used to bring all of Earth's fresh water to a boiling temperature of 100.0°C, after which the water is completely vaporized.

a. How much energy must be added to the fresh water through heat in order to raise its temperature to the boiling point and vaporize it?

b. If the rate at which energy is transferred from the sun is 4.0×10^{26} J/s, how long will it take for the sun to provide sufficient energy for the heating and vaporization process?

NAME ______________________ DATE ______________ CLASS ______________

Thermodynamics

Problem A

WORK DONE ON OR BY A GAS

PROBLEM

Suppose 2.4×10^8 J of work was required to inflate 1.6 million balloons that were released at one time in 1994. If each balloon was filled at a constant pressure that was 25 kPa in excess of atmospheric pressure, what was the change in volume for each balloon?

SOLUTION

Given: $W = 2.4 \times 10^8$ J

$P = 25$ kPa $= 2.5 \times 10^4$ Pa

$n =$ number of balloons $= 1.6 \times 10^6$

Unknown: $\Delta V = ?$

Use the definition of work in terms of changing volume. The total volume change involves 1.6 million balloons, so the volume change of one balloon must be multiplied by $n = 1.6 \times 10^6$.

$$W = nP\Delta V$$

$$\Delta V = \frac{W}{nP} = \frac{2.4 \times 10^8 \text{ J}}{(1.6 \times 10^6)(2.5 \times 10^4 \text{ N/m}^2)} = \boxed{6.0 \times 10^{-3} \text{ m}^3}$$

ADDITIONAL PRACTICE

1. The largest glass bottle made by the method of glass-blowing was over 2 m tall. Suppose the net pressure used to expand the bottle to full volume was 5.1 kPa. If 3.6×10^3 J of work was done in expanding the bottle from an initial volume of 0.0 m^3, what was the final volume?
2. Russell Bradley carried 207 kg of bricks 3.65 m up a ladder. If the amount of work required to perform that task is used to compress a gas at a constant pressure of 1.8×10^6 Pa, what is the change in volume of the gas?
3. Nicholas Mason inflated a weather balloon using just the power of his lungs. The balloon's final radius was 1.22 m. If 642 kJ of work was done to inflate the balloon, at what net pressure was the balloon inflated?
4. Calculate the pressure needed to inflate a sphere that has a volume equal to that of the sun assuming that the work done was 3.6×10^{34} J. The sun's radius is 7.0×10^5 km.
5. In 1979, an extremely low pressure of 87 kPa was measured in a storm over the Pacific Ocean. Suppose a gas is compressed at this pressure and its volume decreases by 25.0×10^{-3} m^3. How much work is done *by* the gas?
6. Susan Williams, of California, blew a bubble-gum bubble with a radius of 29.2 cm. If this were done with a constant net pressure of 25.0 kPa, the work done could have been used to launch a model airplane. If the airplane's mass were 160.0 g, what would have been the launch speed?

NAME ______________________ DATE ____________ CLASS ______________

Thermodynamics

Problem B

THE FIRST LAW OF THERMODYNAMICS

PROBLEM

In 1992, residents of Arkansas consumed, on average, 11.4 L of gasoline per vehicle per day. If this amount of gasoline burns completely in a pure combustion reaction, it will release 4.3×10^8 J of energy. Suppose this amount of energy is transferred by heat from a quantity of gas confined in a very large cylinder. The cylinder, however, is equipped with a piston, and shortly after the energy is transferred by heat from the cylinder, work is done on the gas. The magnitude of the energy transferred by work is equal to one-third the magnitude of the energy transferred by heat. If the initial internal energy of the gas is 1.00×10^9 J, what is the final internal energy of the gas?

SOLUTION

1. DEFINE **Given:** $U_i = 1.00 \times 10^9$ J $\quad Q = -4.3 \times 10^8$ J

$W = Q/3 = -(4.3 \times 10^8 \text{ J})/3 = -1.4 \times 10^8$ J

Work is done on the gas, so work, W, has a negative value and increases the internal energy of the gas. Energy is transferred from the gas by heat, Q, which reduces the gas' internal energy. Therefore, Q must have a negative value.

Unknown: $U_f = ?$

2. PLAN **Choose the equation(s) or situation:** Apply the first law of thermodynamics using the definition of the change in internal energy ($\Delta U = U_f - U_i$) and the values for the energy transferred by heat, Q, and work, W, to find the value for the final internal energy.

$$\Delta U = Q - W$$
$$U_f - U_i = Q - W$$

Rearrange the equation(s) to isolate the unknown(s):

$$U_f = Q - W + U_i$$

3. CALCULATE **Substitute the values into the equation(s) and solve:**

$$U_f = (-4.3 \times 10^8 \text{ J}) - (-1.4 \times 10^8 \text{ J}) + (1.00 \times 10^9 \text{ J})$$
$$U_f = \boxed{7.1 \times 10^8 \text{ J}}$$

4. EVALUATE The final internal energy is less than the initial internal energy because three times as much energy was transferred away from the gas by heat as was transferred to the gas by work done on the gas.

ADDITIONAL PRACTICE

1. The heaviest snake ever found had a mass of 227 kg and measured 8.45 m in length. Suppose a sample of a gas with an initial internal energy of 42.0 kJ performs an amount of work equal to that needed to lift the snake to a height equal to its length. If 4.00 kJ of energy is transferred

to the gas by heat during the lifting process, what will be the final internal energy of the gas?

2. The most massive cannon ever built was made in Russia in 1868. This cannon had a mass in excess of 1.40×10^5 kg and could fire cannonballs with masses up to 4.80×10^2 kg. If the gunpowder in the cannon burned very quickly (adiabatically), forming compressed gas in the barrel, then the gas would expand and perform work on the cannonball. Suppose the initial speed of one of these cannonballs is 2.00×10^2 m/s.

a. How much work is done by the expanding gas?

b. If the internal energy of the gas is 12.0 MJ when the cannonball leaves the barrel, what is the initial internal energy of the gas immediately after all of the powder burns?

3. The world's largest jelly, which had a mass of about 4.00×10^4 kg, was made in Australia in 1981.

a. How much energy must be transferred by heat from the jelly in order for its temperature to decrease 20.0°C? Assume that the specific heat capacity of the jelly equals that of water.

b. Suppose all this energy is transferred by heat to a sample of gas. At the same time, the gas does 1.64×10^9 J of work on its surroundings. What is the net change in the internal energy of the gas?

4. The surface of Lake Ontario is 75.0 m above sea level. The mass of water in the lake is about 1.64×10^{15} kg. Imagine an enormous sample of gas that performs an amount of work capable of lifting Lake Ontario 75.0 m and that also transfers to the lake enough energy by heat to vaporize the entire lake. If the initial temperature of the lake water is 6.0°C and the internal energy of the gas decreases by 90.0 percent, what is the final internal energy of the gas?

5. An average elephant has a mass of 5.00×10^3 kg. Contrary to popular belief, elephants are not slow; they can achieve speeds of up to 40.0 km/h. Imagine a sample of gas that does an amount of work equal to the work required for an average elephant to move from rest to its maximum speed. If the initial internal energy of the gas, 2.50×10^5 J, is to be doubled, how much energy must be transferred to the gas by heat?

6. The rate of nuclear energy production in the United States in 1992 was about 5.9×10^9 J/s. Suppose one second's worth of this energy is transferred by heat to an ideal gas. How much work must be done on or by this gas so that the net increase in its internal energy is 2.6×10^9 J?

7. Dan Koko, a professional stuntman, jumped onto an air pad from a height of almost 1.00×10^2 m. His impact speed was about 141 km/h.

a. If all of Koko's kinetic energy was transferred by heat to the air in the air pad after the inelastic collision with the air pad, how much work was done?

b. Assuming that the initial internal energy of the air in the pad is 40.0 MJ, determine the percent increase in the internal energy of this air after Koko's jump. Assume that Koko's mass was 76.0 kg.

Thermodynamics

Problem C

HEAT-ENGINE EFFICIENCY

PROBLEM

In 1989, Brendan Keenoy ran up 1760 steps in the CN Tower in Toronto, reaching a height of 342 m in 7 min 52 s. Suppose the amount of work done by Keenoy is done by a heat engine. The engine's input energy is 1.34 MJ, and its efficiency is 0.18. How much energy is exhausted from, and how much work is done by the engine?

SOLUTION

Given: $Q_h = 1.34 \text{ MJ} = 1.34 \times 10^6 \text{ J}$ $\quad eff = 0.18$

Unknown: $Q_c = ?$ $\quad W_{net} = ?$

Use the equation for the efficiency of a heat engine, expressed in terms of Q_h and W_{net}. The net work equals the work done in climbing the tower (mgh).

$$Q_c = Q_h(1 - eff) = (1.34 \times 10^6 \text{ J})(1 - 0.18) = (1.34 \times 10^6 \text{ J})(0.82)$$

$$Q_c = \boxed{1.1 \times 10^6 \text{ J}}$$

$$W_{net} = Q_h - Q_c = 1.34 \times 10^6 \text{ J} - 1.1 \times 10^6 \text{ J} = \boxed{2.4 \times 10^5 \text{ J}}$$

ADDITIONAL PRACTICE

1. The oldest working steam engine was designed in 1779 by James Watt. Suppose this engine's efficiency is 8.0 percent. How much energy must be transferred by heat to the engine's surroundings if 2.5 kJ is transferred by heat into the engine? How much work is done?
2. In 1894, the first turbine-driven ship was designed. Suppose the ships turbine's had an efficiency of 16 percent. How much energy would have been exhausted by the turbines if the input energy was 2.0×10^9 J and the net power was 1.5 MW?
3. A steam engine built in 1812 still works at its original site in England. The engine delivers 19 kW of net power. If the engine's efficiency is 6.0 percent, how much energy must be transferred by heat to the engine in 1.00 h?
4. The first motorcycle was built in Germany in 1885, and it could reach a speed of 19 km/h. The output power of the engine was 370 W. If the engine's efficiency was 0.19, what was the input energy after 1.00 min?
5. A fairly efficient steam engine was built in 1840. It required burning only 0.80 kg of coal to perform 2.6 MJ of net work. Calculate the engine's efficiency if burning coal releases 32.6 MJ of energy per kilogram of coal.
6. The world's tallest mobile crane can lift 3.00×10^4 kg to a height of 1.60×10^2 m. What is the efficiency of a heat engine that does the same task while losing 3.60×10^8 J of energy by heat to the surroundings?

NAME ______________________ DATE ____________ CLASS ______________

Vibrations and Waves

Problem A

HOOKE'S LAW

PROBLEM

The pygmy shrew has an average mass of 2.0 g. If 49 of these shrews are placed on a spring scale with a spring constant of 24 N/m, what is the spring's displacement?

SOLUTION

1. DEFINE **Given:**

m = mass of one shrew = 2.0 g = 2.0×10^{-3} kg

$n = 49$

$g = 9.81\ \text{m/s}^2$

$k = 24$ N/m

Unknown:

2. PLAN **Choose the equation(s) or situation:** When the shrews are attached to the spring, the equilibrium position changes. At the new equilibrium position, the net force acting on the shrews is zero. So the spring force (given by Hooke's law) must be equal to and opposite the weight of the shrews.

$$\mathbf{F_{net}} = 0 = \mathbf{F_{elastic}} + \mathbf{F_g}$$

$$F_{elastic} = -kx$$

$$F_g = -m_{tot}g = -nmg$$

$$-kx - nmg = 0$$

Rearrange the equation(s) to isolate the unknown(s):

$$x = \frac{-nmg}{k}$$

3. CALCULATE **Substitute the values into the equation(s) and solve:**

$$x = \frac{-(49)(2.0 \times 10^{-3}\ \text{kg})(9.81\ \text{m/s}^2)}{(24\ \text{N/m})}$$

$$x = \boxed{-4.0 \times 10^{-2}\ \text{m}}$$

4. EVALUATE Forty-nine shrews of 2.0 g each provide a total mass of about 0.1 kg, or a weight of just under 1 N. From the value of the spring constant, a force of 1 N should displace the spring by 1/24 of a meter, or about 4 cm. This indicates that the final result is consistent with the rest of the data.

ADDITIONAL PRACTICE

1. The largest meteorite of lunar origin reportedly has a mass of 19 g. If the meteorite placed on a scale whose spring constant is 83 N/m, what is the compression of the spring?

2. In 1952, a great rainfall hit the island of Reunion in the Indian Ocean. In less than 24 h, 187 kg of rain fell on each square meter of soil. If a 187 kg mass is placed on a scale that has a spring constant of 1.53×10^4 N/m, how far is the spring compressed?

3. The largest tigers, and therefore the largest members of the cat family, are the Siberian tigers. Male Siberian tigers are reported to have an average mass of about 389 kg. By contrast, a variety of very small cat that is native to India has an average adult mass of only 1.5 kg. Suppose this small cat is placed on a spring scale, causing the spring to be extended from its equilibrium position by 1.2 mm. How far would the spring be extended if a typical male Siberian tiger were placed on the same scale?

4. The largest known crab is a giant spider crab that had a mass of 18.6 kg. The distance from the end of one of this crab's claws to the end of the other claw measured about 3.7 m. If this particular giant spider crab were hung from an elastic band so that the elongation of the band was equal to the crab's claw span, what would be the spring constant of the elastic band?

5. The CN Tower in Toronto, Canada, is 533 m tall, making it the world's tallest free-standing structure. Suppose an unusually long bungee cord is attached to the top of the CN Tower. The equilibrium length of the cord is equal to one-third the height of the tower. When a test mass of 70.0 kg is attached, the cord stretches to a length that equals two-thirds of the tower's height. From this information, determine the spring constant of the bungee cord.

6. The largest ruby in the world may be found in New York. This ruby is 109 mm long, 91 mm wide, and 58 mm thick, making its volume about 575 cm^3. (By comparison, the world's largest diamond, the Star of Africa, has a volume of just over 30 cm^3.)

a. If the ruby is attached to a vertically hanging spring with a spring constant of 2.00×10^2 N/m so that the spring is stretched 15.8 cm what is the gravitational force pulling the spring?

b. What is the mass of the jewel?

7. Mauna Kea on the island of Hawaii stands 4200 m above sea level. However, when measured from the island's sea-submerged base, Mauna Kea has a height of 10 200 m, making it the tallest single mountain in the world. If you have a 4.20×10^3 m elastic cord with a spring constant of 3.20×10^{-2} N/m, what force would stretch the spring to 1.02×10^4 m?

8. Rising 348 m above the ground, La Gran Piedra in Cuba is the tallest rock on Earth. Suppose an elastic band 2.00×10^2 m long hangs vertically off the top of La Gran Piedra. If the band's spring constant is 25.0 N/m, how large must a mass be if, when it is attached to the band, it causes the band to stretch all the way to the ground?

Vibrations and Waves

Problem B

SIMPLE HARMONIC MOTION OF A SIMPLE PENDULUM

PROBLEM

Two friends in France use a pendulum hanging from the world's highest railroad bridge to exchange messages across a river. One friend attaches a letter to the end of the pendulum and releases it so that the pendulum swings across the river to the other friend. The bridge is 130.0 m above the river. How much time is needed for the letter to make one swing across the river? Assume the river is 16.0 m wide.

SOLUTION

Given: $L = 130.0$ m $\quad a_g = g = 9.81$ m/s^2

Unknown: t = time required for pendulum to cross river = $T/2$ = ?

Use the equation for the period of a simple pendulum. Then divide the period by two to find the time of one swing across the river. The width of the river is not needed to calculate the answer, but it must be small compared to the length of the pendulum in order to use the equations for simple harmonic motion.

$$T = 2\pi\sqrt{\frac{L}{a_g}} = 2\pi\sqrt{\frac{130.0 \text{ m}}{9.81 \text{ m/s}^2}} = 22.9 \text{ s}$$

$$t = \frac{T}{2} = \frac{22.9 \text{ s}}{2} = \boxed{11.4 \text{ s}}$$

ADDITIONAL PRACTICE

1. An earthworm found in Africa was 6.7 m long. If this worm were a simple pendulum, what would its period be?
2. The shortest venomous snake, the spotted dwarf adder, has an average length of 20.0 cm. Suppose this snake hangs by its tail from a branch and holds a heavy prey with its jaws, simulating a pendulum with a length of 15.0 cm. How long will it take the snake to swing through one period?
3. If bamboo, which can grow 88 cm in a day, is grown for four days and then used to make a simple pendulum, what will be the pendulum's period?
4. A simple pendulum with a frequency of 6.4×10^{-2} Hz is as long as the largest known specimen of Pacific giant seaweed. What is this length?
5. The deepest permafrost is found in Siberia, Russia. Suppose a shaft is drilled to the bottom of the frozen layer, and a simple pendulum with a length equal to the depth of the shaft oscillates within the shaft. In 1.00 h the pendulum makes 48 oscillations. Find the depth of the permafrost.
6. Ganymede, the largest of Jupiter's moons, is also the largest satellite in the solar system. Find the acceleration of gravity on Ganymede if a simple pendulum with a length of 1.00 m has a period of 10.5 s.

NAME ______________________ DATE ______________ CLASS ______________

Vibrations and Waves

Problem C

SIMPLE HARMONIC MOTION OF A MASS-SPRING SYSTEM

PROBLEM

A large pearl was found in the Philippines in 1934. Suppose the pearl is placed on a spring scale whose spring constant is 362 N/m. If the scale's platform oscillates with a frequency of 1.20 Hz, what is the mass of the pearl?

SOLUTION

Given: $k = 362$ N/m $\quad f = 1.20$ Hz

Unknown: $m = ?$

Use the equation for the period of a mass-spring system. Then express the period in terms of frequency ($T = 1/f$).

$$T = 2\pi\sqrt{\frac{m}{k}} = \frac{1}{f}$$

$$m = \frac{k}{4\pi^2 f^2} = \frac{362 \text{ N/m}}{4\pi^2(1.20 \text{ Hz})^2} = \boxed{6.37 \text{ kg}}$$

ADDITIONAL PRACTICE

1. The hummingbird makes a humming sound with its wings, which beat with a frequency of 90.0 Hz. Suppose a mass is attached to a spring with a spring constant of 2.50×10^2 N/m. How large is the mass if its oscillation frequency is 3.00×10^{-2} times that of a hummingbird's wings?
2. In 1986, a 35×10^3 kg watch was demonstrated in Canada. Suppose this watch is placed on a huge trailer that rests on a lightweight platform, and that oscillations equal to 0.71 Hz are induced. Find the trailer's mass if the platform acts like a spring scale with a spring constant equal to 1.0×10^6 N/m.
3. A double coconut can grow for 10 years and have a mass of 20.0 kg. If a 20.0 kg double coconut oscillates on a spring 42.7 times each minute, what is the spring constant of the spring?
4. The monument commemorating the Battle of San Jacinto in Texas stands almost 2.00×10^2 m and is topped by a 2.00×10^5 kg star. Imagine that a 2.00×10^5 kg mass is placed on a spring platform. The platform requires 0.80 s to oscillate from the top to the bottom positions. What is the spring constant of the spring supporting the platform?
5. Suppose a 2662 kg giant seal is placed on a scale and produces a 20.0 cm compression. If the seal and spring system are set into simple harmonic motion, what is the period of the oscillations?
6. On average, a newborn human's mass is just over 3.0 kg. However, in 1955, a 10.2 kg boy was born in Italy. Suppose this baby is placed in a crib hanging from springs with a total spring constant of 2.60×10^2 N/m. If the cradle is rocked with simple harmonic motion, what is its period?

Vibrations and Waves

Problem D

WAVE SPEED

PROBLEM

The world's largest guitar, which was built by high school students in Indiana, has strings that are 9.0 m long. The fundamental vibration that can be induced on each string has a wavelength equal to twice the string's length. If the wave speed in a string is 9.0×10^2 m/s, what is the frequency of vibration?

SOLUTION

Given: $f = 50.0$ Hz $L = 9.0$ m

Unknown: $v = ?$

Use the equation for the speed of a wave. The wavelength is equal to twice the length of the string ($\lambda = 2L$).

$$v = f\lambda = f(2L) = (50.0\text{ Hz})[(2)(9.0\text{ m})] = \boxed{9.0 \times 10^2\text{ m/s}}$$

ADDITIONAL PRACTICE

1. The speed of sound in sea water is about 1530 m/s. If a sound wave has a frequency of 2.50×10^2 Hz, what is its wavelength in sea water?
2. Cicadas produce a sound that has a frequency of 123 Hz. What is the wavelength of this sound in the air? The speed of sound in air is 334 m/s.
3. Human fingers are very sensitive, detecting vibrations with amplitudes as low as 2.0×10^{-5} m. Consider a sound wave with a wavelength exactly 1000 times greater than the lowest amplitude detectable by fingers. What is this wave's frequency?
4. A nineteenth-century fisherman's cottage in England is only 2.54 m long. Suppose a fisherman whistles inside the cottage, producing a note that has a wavelength that exactly matches the length of the house. What is the whistle's frequency? The speed of sound in air is 334 m/s.
5. The lowest vocal note in the classical repertoire is low D ($f = 73.4$ Hz), which occurs in an aria in Mozart's opera *Die Entführung aus dem Serail.* If low D has a wavelength of 4.50 m, what is the speed of sound in air?
6. The highest-pitched sound that a human ear can detect is about 21 kHz. On the other hand, dolphins can hear ultrasound with frequencies up to 280 kHz. What is the speed of sound in water if the wavelength of ultrasound with a frequency of 2.80×10^5 Hz is 0.510 cm? How long would it take this sound wave to travel to a dolphin 3.00 km away?

NAME ______________________ DATE ____________ CLASS ______________

Sound

Problem A

INTENSITY OF SOUND WAVES

PROBLEM

Kåre Walkert of Sweden reportedly snores loudly, with a record intensity of 4.5×10^{-8} W/m^2. Suppose the intensity of Walkert's snores are measured 0.60 m from her mouth. What is the power associated with the record snore?

SOLUTION

Given: Intensity $= 4.5 \times 10^{-8}$ W/m^2

$r = 0.60$ m

Unknown: $P = ?$

Use the equation for the intensity of a spherical wave.

$$\text{Intensity} = \frac{P}{4\pi r^2}$$

$$P = 4\pi r^2(\text{Intensity}) = 4\pi(0.60\text{ m})^2(4.5 \times 10^{-8}\text{ W/m}^2)$$

$$P = \boxed{2.0 \times 10^{-7}\text{ W}}$$

ADDITIONAL PRACTICE

1. Blue whales are the loudest creatures; they can emit sound waves with an intensity of 3.0×10^{-3} W/m^2. If this intensity is measured 4.0 m from its source, what power is associated with the sound wave?
2. The whistling sound that is characteristic of the language known as "silbo," which is used on the Canary Island of Gomera, is detectable at 8.0 km. Use the spherical wave approximation to find the power of a whistler's sound. Sound intensity at the hearing threshold is 1.0×10^{-12} W/m^2.
3. Estimate how far away a cicada can be heard if the lowest audible intensity of the sound it produces is 1.0×10^{-12} W/m^2 and the power of a cicada's sound source is 2.0×10^{-6} W.
4. Howler monkeys, found in Central and South America, can emit a sound that can be heard by a human several miles away. The power associated with the sound is roughly 3.0×10^{-4} W. If the threshold of hearing of a human is assumed to be 1.1×10^{-13} W/m^2, how far away can a howler monkey be heard.
5. In 1983, Roy Lomas became the world's loudest whistler; the power of his whistle was 1.0×10^{-4} W. What was the sound's intensity at 2.5 m?
6. In 1988, Simon Robinson produced a sound having an intensity level of 2.5×10^{-6} W/m^2 at a distance of 2.5 m. What power was associated with Robinson's scream?

Sound

Problem B

HARMONICS

PROBLEM

The tallest load-bearing columns are part of the Temple of Amun in Egypt, built in 1270 B.C. Find the height of these columns if a standing wave with a frequency of 47.8 Hz is generated in an open pipe that is as tall as the columns. The sixth harmonic is generated. The speed of sound in air is 334 m/s.

SOLUTION

Given: $f_6 = 47.8$ Hz $\quad v = 334$ m/s

n = number of the harmonic = 6

Unknown: $L = ?$

When the pipe is open, the wavelength associated with the first harmonic (fundamental frequency) is twice the length of the pipe.

$$f_n = n\frac{v}{2L} \qquad n = 1, 2, 3, \ldots$$

$$L = n\frac{v}{2f_n} = (6)\frac{(334\text{m/s})}{2(47.8\text{ Hz})} = \boxed{21.0\text{ m}}$$

ADDITIONAL PRACTICE

1. A 47.0 m alphorn was made in Idaho in 1989. An alphorn behaves like a pipe with one end closed. If the frequency of the fifteenth harmonic is 26.7 Hz, how long is the alphorn? The speed of sound in air is 334 m/s.
2. A fully functional acoustic guitar over 8.0 m in length is on display in Bristol, England. Suppose the speed of waves on the guitar's strings is 5.00×10^2 m/s. If a third harmonic is generated on a string, so that the sound produced in air has a wavelength of 3.47 m, what is the length of the string? The speed of sound in air is 334 m/s.
3. The unsupported flagpole built for Canada's Expo 86 has a height of 86 m. If a standing wave with a 19th harmonic is produced in an 86 m open pipe, what is its frequency? The speed of sound in air is 334 m/s.
4. A power-plant chimney in Spain is 3.50×10^2 m high. If a standing wave with a frequency of 35.5 Hz is generated in an open pipe with a length equal to the chimney's height and the 75th harmonic is present, what is the speed of sound?
5. The world's largest organ was completed in 1930 in Atlantic City, New Jersey. Its shortest pipe is 4.7 mm long. If one end of this pipe is closed, what is the number of harmonics created by an ultrasound with a wavelength of 3.76 mm?

NAME ______________________ DATE ____________ CLASS ______________

Light and Reflection

Problem A

ELECTROMAGNETIC WAVES

PROBLEM

The atoms in an HCl molecule vibrate like two charged balls attached to the ends of a spring. If the wavelength of the emitted electromagnetic wave is 3.75 μm, what is the frequency of the vibrations?

SOLUTION

Given: $\lambda = 3.75 \times 10^{-6}$ m

$c = 3.00 \times 10^{8}$ m/s

Unknown: $f = ?$

Use the wave speed equation, and solve for λ.

$$c = f\lambda$$

$$f = \frac{c}{\lambda} = \frac{3.00 \times 10^{8}\ \text{m/s}}{3.75 \times 10^{-6}\ \text{m}} = \boxed{8.00 \times 10^{13}\ \text{Hz}}$$

ADDITIONAL PRACTICE

1. New-generation cordless phones use a 9.00×10^{2} MHz frequency and can be operated up to 60.0 m from their base. How many wavelengths of the electromagnetic waves can fit between your ear and a base 60.0 m away?
2. The highest *directly* measured frequency is 5.20×10^{14} Hz, corresponding to one of the transitions in iodine-127. How many wavelengths of electromagnetic waves with this frequency could fit across a dot on a book page? Assume the dot is 2.00×10^{-4} m in diameter.
3. Commercial trucks cause about 18 000 lane-change and merging accidents per year in the United States. To prevent many of them, a warning system covering blind spots is being developed. The system uses electromagnetic waves of frequency 2.40×10^{10} Hz. What is the wavelength of these waves?
4. A typical compact disc stores information in tiny pits on the disc's surface. A typical pit size is 1.2 μm. What is the frequency of electromagnetic waves that have a wavelength equal to the typical CD pit size?
5. A new antiterrorist technique detects the differences in electromagnetic waves emitted by humans and by weapons made of metal, plastic, or ceramic. One possible range of wavelengths used with this technique is from 2.0 mm to 5.0 mm. Calculate the associated range of frequencies.
6. The U.S. Army's loudest loudspeaker is almost 17 m across and is transported on a special trailer. The sound is produced by an electromagnetic coil that can generate a minimum frequency of 10.0 Hz. What is the wavelength of these electromagnetic waves?

NAME ______________________ DATE ____________ CLASS ____________

Light and Reflection

Problem B

IMAGING WITH CONCAVE MIRRORS

PROBLEM

Lord Rosse, who lived in Ireland in the nineteenth century, built a reflecting telescope called the Leviathan. Lord Rosse used it for astronomical observations and discovered the spiral form of galaxies. Suppose the Leviathan's mirror has a focal length of 2.50 m. Where would you place an object in front of the mirror in order to form an image at a distance of 3.75 m? What would the magnification be? If the image height were 6.0 cm, what would the object height be?

SOLUTION

1. DEFINE

Given: $f = +2.50$ m $\qquad q = +3.75$ m

$h = 6.0$ cm

The mirror is concave, so f is positive. The object is in front of the mirror, so q is positive.

Unknown: $p = ?$ $\qquad M = ?$

Diagram:

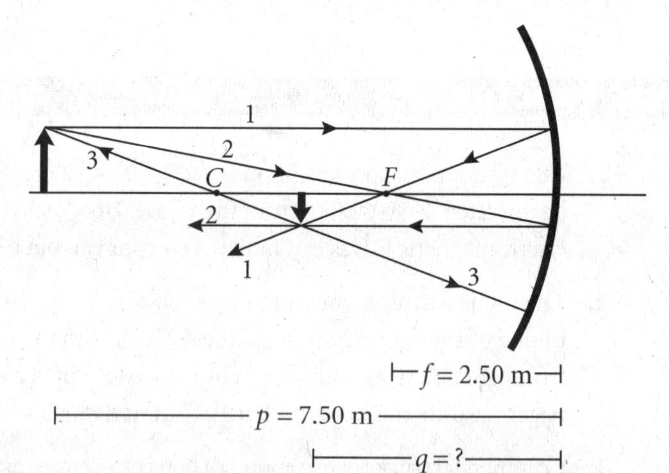

2. PLAN

Choose the equation(s) or situation: Use the mirror equation for focal length and the magnification formula.

$$\frac{1}{p} + \frac{1}{q} = \frac{1}{f} \qquad M = -\frac{q}{p}$$

Rearrange the equation(s) to isolate the unknown(s):

$$\frac{1}{p} = \frac{1}{f} - \frac{1}{q}$$

3. CALCULATE

Substitute the values into the equation(s) and solve:

$$\frac{1}{p} = \frac{1}{2.50 \text{ m}} - \frac{1}{3.75 \text{ m}} = \frac{0.400}{1 \text{ m}} - \frac{0.267}{1 \text{ m}} = \frac{0.133}{1 \text{ m}}$$

$$p = \boxed{7.50 \text{ m}}$$

4. EVALUATE Substitute the values for p and q to find the magnification of the image and h' to find the object height.

$$M = -\frac{3.75\text{ m}}{7.50\text{ m}} = \boxed{-0.500}$$

$$h = -\frac{ph'}{q} = -\frac{(7.50\text{ m})(0.060\text{ m})}{3.75\text{ m}} = \boxed{0.12\text{ m}}$$

The image appears between the focal point (2.50 m) and the center of curvature, is smaller than the object, and is inverted ($-1 < M < 0$). These results are confirmed by the ray diagram. The image is therefore real.

ADDITIONAL PRACTICE

1. In Alaska, the top of Mount McKinley has been seen from the top of Mount Sanford, a distance of 370 km. An object is 3.70×10^2 km from a giant concave mirror. If the focal length of the mirror is 2.50×10^2 km what are the object distance and the magnification?

2. A human hair is about 80.0 μm thick. If one uses a concave mirror with a focal length of 2.50 cm and obtains an image of −59.0 cm, how far has the hair been placed from the mirror? What is the magnification of the hair?

3. A mature blue whale may have a length of 28.0 m. How far from a concave mirror with a focal length of 30.0 m must a 7.00-m-long baby blue whale be placed to get a real image the size of a mature blue whale?

4. In 1950 in Seattle, Washington, there was a Christmas tree 67.4 m tall. How far from a concave mirror having a radius of curvature equal to 12.0 m must a person 1.69 m tall stand to form a virtual image equal to the height of the tree? Will the image be upright or inverted?

5. A stalagmite that is 32 m tall can be found in a cave in Slovakia. If a concave mirror with a focal length of 120 m is placed 180 m from this stalagmite, how far from the mirror will the image form? What is the size of the image? Is it upright or inverted? real or virtual?

6. The eye of the Atlantic giant squid has a diameter of 5.00×10^2 mm. If the eye is viewed in a concave mirror with a radius of curvature equal to the diameter of the eye and the eye is 1.000×10^3 mm from the mirror, how far is the image from the mirror? What is the size of the image? Is the image real or virtual?

7. *Quick Bird* is the first commercial satellite designed for forming high-resolution images of objects on Earth. Suppose the satellite is 1.00×10^2 km above the ground and uses a concave mirror to form a primary image of a 1.00 m object. If the image size is 4.00 μm and the image is inverted, what is the mirror's radius of curvature?

8. A stalactite with a length of 10.0 m was found in Brazil. If the stalactite is placed 18.0 m in front of a concave mirror, a real image 24.0 m tall is formed. Calculate the mirror's radius of curvature.

Light and Reflection

Problem C

CONVEX MIRRORS

PROBLEM

The largest jellyfish ever caught had tentacles up to 36 m long, which is greater than the length of a blue whale. Suppose the jellyfish is located in front of a convex spherical mirror 36.0 m away. If the mirror has a focal length of 12.0 m, how far from the mirror is the image? What is the image height of the jellyfish?

SOLUTION

1. DEFINE

Given: $f = -12.0$ m $\quad p = +36.0$ m

$h = 36$ m

The mirror is convex, so f is negative. The object is in front of the mirror, so p is positive.

Unknown: $q = ?$ $\quad M = ?$

Diagram:

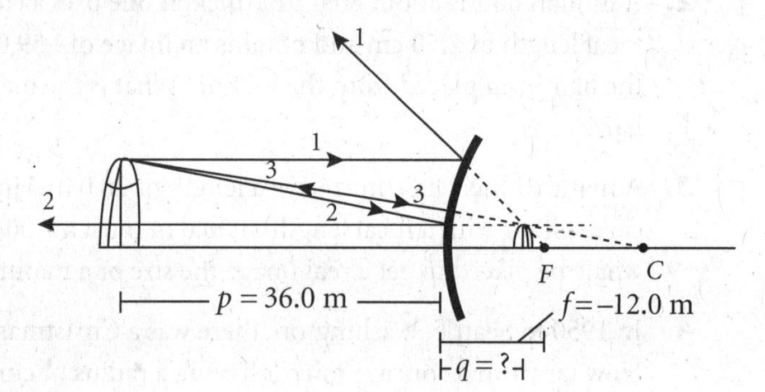

2. PLAN

Choose the equation(s) or situation: Use the mirror equation for focal length and the magnification formula.

$$\frac{1}{p} + \frac{1}{q} = \frac{1}{f} \qquad M = -\frac{q}{p}$$

Rearrange the equation(s) to isolate the unknown(s):

$$\frac{1}{q} = \frac{1}{f} - \frac{1}{p}$$

Substitute the values into the equation(s) and solve:

3. CALCULATE

$$\frac{1}{q} = -\frac{1}{12.0\text{ m}} - \frac{1}{36.0\text{ m}} = -\frac{0.0833}{1\text{ m}} - \frac{0.0278}{1\text{ m}} = -\frac{0.1111}{1\text{ m}}$$

$$q = \boxed{-9.001\text{ m}}$$

Substitute the values for p and q to find the magnification of the image and h to find the image height.

$$M = -\frac{-9.001\text{ m}}{36.0\text{ m}} = \boxed{0.250}$$

$$h' = -\frac{qh}{p} = -\frac{(-9.001\text{ m})(36\text{ m})}{(36.0\text{ m})} = \boxed{9.001\text{ m}}$$

The image appears between the focal point (–12.0 m) and the mirror's surface, as confirmed by the ray diagram. The image is smaller than the object ($M < 1$) and is upright ($M > 0$), as is also confirmed by the ray diagram.

ADDITIONAL PRACTICE

1. The radius of Earth is 6.40×10^3 km. The moon is about 3.84×10^5 km away from Earth and has a diameter of 3475 km. The Pacific Ocean surface, which can be considered a convex mirror, forms a virtual image of the moon. What is the diameter of that image?

2. A 10 g thread of wool was produced by Julitha Barber of Australia in 1989. Its length was 553 m. Suppose Barber is standing a distance equal to the thread's length from a convex mirror. If the mirror's radius of curvature is 1.20×10^2 m, what will the magnification of the image be?

3. Among the many discoveries made with the Hubble Space Telescope are four new moons of Saturn, the largest being just about 70.0 km in diameter. Suppose this moon is covered by a highly reflective coating, thus forming a spherical convex mirror. Another moon happens to pass by at a distance of 1.00×10^2 km. What is the image distance?

4. The largest scale model of the solar system was built in Peoria, Illinois. In this model the sun has a diameter of 11.0 m. The real diameter of the sun is 1.4×10^6 km. What is the scale to which the sun's size has been reduced in the model? If the model's sun is a reflecting sphere, where in front of the sphere is the object located?

5. Bob Henderson of Canada built a model railway to a scale of 1:1400. How far from a convex mirror with a focal length of 20.0 mm should a full-size engine be placed so that the size of its virtual image is the same as that of the model engine?

6. The largest starfish ever discovered had a diameter of 1.38 m. Suppose an object of this size is placed 6.00 m in front of a convex mirror. If the image formed is just 0.900 cm in diameter (the size of the smallest starfish), what is the radius of curvature of the mirror?

7. In 1995, a functioning replica of the 1936 Toyota Model AA sedan was made in Japan. The model is a mere 4.78 mm in length. Suppose an object measuring 12.8 cm is placed in front of a convex mirror with a focal length of 64.0 cm. If the size of the image is the same as the size of the model car, how far is the image from the mirror's surface?

8. Some New Guinea butterflies have a wingspan of about 2.80×10^2 mm. However, some butterflies which inhabit the Canary Islands have a wingspan of only 2.00 mm. Suppose a butterfly from New Guinea is placed in front a convex mirror. The image produced is the size of a butterfly from the Canary Islands. If the image is 50.0 cm from the mirror's surface, what is the focal length of the mirror?

NAME ______________________ DATE ______________ CLASS ______________

Refraction

Problem A

SNELL'S LAW

PROBLEM

The smallest brilliant-cut diamond has a mass of about 15 μg and a height of just 0.11 mm. Suppose a ray of light enters the diamond from the air and, upon contact with one of the gem's facets, refracts at an angle of 22.2°. If the angle of incidence is 65.0°, what is the diamond's index of refraction?

SOLUTION

Given: $\theta_i = 65.0°$ $\theta_r = 22.2°$ $n_i = 1.00$

Unknown: $n_r = ?$

Use the equation for Snell's law.

$$n_i \sin \theta_i = n_r \sin \theta_r$$

$$n_r = n_i \frac{\sin \theta_i}{\sin \theta_r} = (1.00)\frac{(\sin 65.0°)}{(\sin 22.2°)} = \boxed{2.40}$$

ADDITIONAL PRACTICE

1. Extra dense flint glass has one of the highest indices of refraction of any type of glass. Suppose a beam of light passes from air into a block of extra dense flint glass. If the light has an angle of incidence of 72° and an angle of refraction of 34°, what is the index of refraction of the glass?

2. The index of refraction of a clear oil is determined by passing a beam of light through the oil and measuring the angles of incidence and refraction. If the light in air approaches the oil's surface at an angle of 47.9° to the normal and moves into the oil at an angle of 29.0° to the normal, what is the oil's index of refraction? Assume the index of refraction for air is 1.00.

3. Someone on a glass-bottom boat shines a light through the glass into the water below. A scuba diver beneath the boat sees the light at an angle of 17° with respect to the normal. If the glass's index of refraction is 1.5 and the water's index of refraction is 1.33, what is the angle of incidence with which the light passes from the glass into the water? What is the angle of incidence with which the light passes from the air into the glass?

4. A beam of light is passed through a layer of ice into a fresh-water lake below. The angle of incidence for the light in the ice is 55.0°, while the angle of refraction for the light in the water is 53.8°. Calculate the index of refraction of the ice, using 1.33 as the index of refraction of fresh water.

5. An arrangement of three glass blocks with indices of refraction of 1.5, 1.6, and 1.7 are sandwiched together. A beam of light enters the first block from air at an angle of 48° with respect to the normal. What is the angle of refraction after the light enters the third block?

Refraction

Problem B

LENSES

PROBLEM

Suppose the smallest car that is officially allowed on United States roads is placed upright in front of a converging lens. The lens, which has a focal length of 1.50 m, forms an image 75.0 cm tall and 2.00 m away. Calculate the object distance, the magnification, and the object height.

SOLUTION

1. DEFINE **Given:** $q = +2.00$ m $\quad f = +1.50$ m $\quad h' = -0.750$ m

The image is behind the lens, so q is positive. The lens is converging, so the focal length is positive ($f > 0$). The image is inverted, so h' is negative.

Unknown: $p = ?$ $\quad M = ?$ $\quad h = ?$

Diagram:

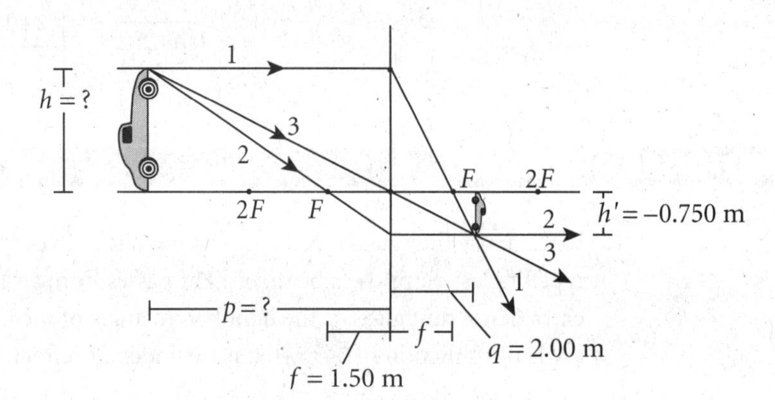

2. PLAN **Choose the equation(s) or situation:** Use the thin-lens equation to calculate the image distance. Then use the equation for magnification to calculate the magnification and the object height.

$$\frac{1}{f} = \frac{1}{p} + \frac{1}{q} \qquad M = -\frac{q}{p} = \frac{h'}{h}$$

Rearrange the equation(s) to isolate the unknown(s):

$$\frac{1}{p} = \frac{1}{f} - \frac{1}{q}$$

3. CALCULATE **Substitute the values into the equation(s) and solve:**

$$\frac{1}{p} = \frac{1}{1.50 \text{ m}} - \frac{1}{2.00 \text{ m}} = \frac{0.667}{1 \text{ m}} - \frac{0.500}{1 \text{ m}} = \frac{0.167}{1 \text{ m}} \qquad p = \boxed{6.00 \text{ m}}$$

$$M = -\frac{q}{p} = -\frac{(2.00 \text{ m})}{(6.00 \text{ m})} = \boxed{-0.333}$$

$$h = \frac{h'}{M} = \frac{(-0.750 \text{ m})}{(-0.333)} = \boxed{2.25 \text{ m}}$$

4. EVALUATE Because $-1 < M < 1$, the image must form at a distance less than $2f$ but greater than f, which is the case. At this position the image is real and inverted.

NAME ______________________ DATE ____________ CLASS ______________

ADDITIONAL PRACTICE

1. The National Museum of Photography, Film & Television, in England, has a huge converging lens with a diameter of 1.37 m and a focal length of 8.45 m. Suppose you use this lens as a magnifying glass. At what distance would a friend have to stand for the friend's image to appear 25 m in front of the lens? What is the image magnification?

2. The largest of seals is the elephant seal, while the smallest seal, the Galápagos fur seal, is only 1.50 m in length. Suppose you use a diverging lens with a focal length of 8.58 m to observe an elephant seal. The elephant seal's image turns out to have the exact length of a Galápagos fur seal and forms 6.00 m in front of the lens. How far away is the elephant seal, and what is its length?

3. The common musk turtle, also called a "stinkpot," has a length of 7.60 cm at maturity. Suppose a turtle with this length is placed in front of a diverging lens that has a 14.0 cm focal length. If the turtle's image is 4.00 cm across, how far is the turtle from the lens? How far is the turtle's image from the lens?

4. The largest mammal on land, the elephant, can reach a height of 3.5 m. The largest mammal in the sea, however, is much bigger. A blue whale, which is also the largest animal ever to have lived on Earth, can be as long as 28 m. If you use a diverging lens with a focal length of 10.0 m to look at a 28.0-m-long blue whale, how far must you be from the whale to see an image equal to an elephant's height (3.50 m)?

5. The largest scorpions in the world live in India. The smallest scorpions live on the shore of the Red Sea and are only about 1.40 cm in length. Suppose a diverging lens with a focal length of 20.0 cm forms an image that is 1.40 cm wide. If the image is 19.00 cm in front of the lens, what is the object distance and size?

6. The ocean sunfish, *Mola mola*, produces up to 30×10^6 eggs at a time. Each egg is about 1.3 mm in diameter. How far from a magnifying glass with a focal length of 6.0 cm should an egg be placed to obtain an image 5.2 mm in size? How far is it between the image and the lens?

7. In 1992, Thomas Bleich of Austin, Texas, produced a photograph negative of about 3500 attendants at a concert. The negative was more than 7 m long. Bleich used a panoramic camera with a lens that had a focal length of 26.7 cm. Suppose this camera is used to take a picture of just one concert attendant. If the attendant is 3.00 m away from the lens, how far should the film be from the lens? What is the image magnification?

8. Komodo dragons, or monitors, are the largest lizards, having an average length of 2.25 m. This is much shorter than the largest crocodiles. If a crocodile is viewed through a diverging lens with a focal length of 5.68 m, its image is 2.25 m long. If the crocodile is 12.0 m from this lens, what is the image distance? How long is the crocodile?

9. The body of the rare thread snake is as thin as a match, and the longest specimen ever found was only 108 mm long. If a thread snake of this length is placed a distance equal to four times its length from a diverging lens and the lens has a focal length of 216 mm, how long is the snake's image? How far from the lens is the image?

10. Tests done by the staff of *Popular Photography* magazine revealed that the zoom lenses available in stores have a focal length different from what is written on them. Suppose one of these lenses, which is identified as having a focal length of 210 mm, yields an upright image of an object located 117 mm away. If the image magnification is 2.4, what is the true focal length of the lens?

11. In 1994, a model car was made at a scale of 1:64. This car traveled more than 600 km in 24 h, setting a record. If this model car is placed under an opaque projector, a real image will be projected. Suppose the image on the screen has the same size as the actual, full-scale car. If the screen is 12 m from the lens, what is the focal length of the lens? Is the image upright or inverted?

12. The tallest man in history, Robert Wadlaw, was 2.72 m tall. The smallest woman in history, Pauline Musters, had a height of 0.55 m. Suppose Wadlaw is 5.0 m away from a converging lens. If his image is the same size as Musters, what is the focal length of the lens?

13. Hummingbirds eggs, which have an average size of 10.0 mm, are the smallest eggs laid by any bird. Suppose an egg is placed 12.0 cm from a magnifying glass. A virtual image with a magnification of 3.0 is produced. What is the focal length of the lens?

14. In 1876, the *Daily Banner*, a newspaper printed in Roseberg, Oregon, had pages that were 7.60 cm wide. What would be the width of this newspaper's image if the newspaper were placed 16.0 cm from a diverging lens with a focal length of 12.0 cm?

15. Estimates show that the largest dinosaurs were 48 m long. Suppose you take a trip back in time with a camera that has a focal length of 110 mm. Coming across a specimen of the largest dinosaur, you take its picture, but to be safe and inconspicuous you take it from a distance of 120 m. What length will the image have on the film?

16. The smallest spiders in the world are only about 0.50 mm across. On the other hand, the goliath tarantula, of South America, can have a leg span of about 280 mm. Suppose you use a diverging lens with a focal length of 0.80 m to obtain an image that is 0.50 mm wide of an object that is 280 mm wide. How far is the object from the lens? How far is the image from the lens?

NAME ______________________ DATE ______________ CLASS ______________

Refraction

Problem C

CRITICAL ANGLE

PROBLEM

Rutile, TiO_2, has one of the highest indices of refraction: 2.80. Suppose the critical angle between rutile and an unknown liquid is 33.6°. What is the liquid's index of refraction?

SOLUTION

Given: $\theta_c = 33.6°$

$n_i = 2.80$

Unknown: $n_r = ?$

Use the equation for critical angle.

$$\sin \theta_c = \frac{n_r}{n_i}$$

$$n_r = n_i \sin \theta_c = (2.80)(\sin 33.6°) = \boxed{1.54}$$

ADDITIONAL PRACTICE

1. Light moves from glass into a substance of unknown refraction index. If the critical angle for the glass is 46° and the index of refraction for the glass is 1.5, what is the index of refraction of the other substance?
2. The largest uncut diamond had a mass of more than 600 g. Eventually, the diamond was cut into several pieces. Suppose one of those pieces is a cube with sides 1.00 cm wide. If a beam of light were to pass from air into the diamond with an angle of incidence equal to 75.0°, the angle of refraction would be 23.3°. From this information, calculate the index of refraction and the critical angle for diamond in air.
3. A British company makes optical fibers that are 13.6 km in length. If the critical angle for the fibers in air is 42.1°, what is the index of refraction of the fiber material?
4. In 1996, the Fiberoptic Link Around the Globe (FLAG) was started. It initially involves placing a 27 000 km fiber optic cable at the bottom of the Mediterranean Sea and the Indian Ocean. Suppose the index of refraction of this fiber is 1.56 and the index of refraction of sea water 1.36, what is critical angle for internal reflection in the fiber?
5. The world's thinnest glass is 0.025 mm thick. If the index of refraction for this glass is 1.52, what is the critical angle of ocean water? How far will a ray of light travel in the glass if it undergoes one internal reflection at the critical angle?

NAME ______________________ DATE ______________ CLASS ______________

Interference and Diffraction

Problem A

INTERFERENCE AND DIFFRACTION

PROBLEM

To help prevent cavities, scientists at the University of Rochester have developed a method for melting tooth enamel without disturbing the inner layers, or pulp, of the tooth. To accomplish this, short pulses from a laser are used. These laser pulses, which are in the microwave portion of the electromagnetic spectrum, have a wavelength of 9.3 μm. Suppose this laser is operated continuously with a double-slit arrangement. If the slits have a separation of 45 μm, at what angle will the third-order maximum be observed?

SOLUTION

1. DEFINE **Given:**

$\lambda = 9.3\ \mu\text{m} = 9.3 \times 10^{-6}\ \text{m}$

$d = 45\ \mu\text{m} = 45 \times 10^{-6}\ \text{m}$

$m = 3$

Unknown: $\theta = ?$

2. PLAN **Choose the equation(s) or situation:** Because a maximum (bright) fringe is observed, the equation for constructive interference should be used.

$$d \sin \theta = m\lambda$$

Rearrange the equation(s) to isolate the unknown(s):

$$\theta = \sin^{-1}\left(\frac{m\lambda}{d}\right)$$

3. CALCULATE **Substitute the values into the equation(s) and solve:**

$$\theta = \sin^{-1}\left(\frac{3(9.3 \times 10^{-6}\ \text{m})}{45 \times 10^{-6}\ \text{m}}\right)$$

$$\theta = \boxed{38°}$$

4. EVALUATE The angle at which the third-order maximum appears is 38° from the central maximum. Although the wavelength of the electromagnetic radiation is large compared with that of visible light, the separation of the slits is much larger than it would be in a visible-light double-slit setup. If the same slit separation that is used with visible light (typically on the order of a few micrometers) were used with microwave radiation, the interference pattern would not appear because even for $m = 1$, $\sin \theta$ would be greater than one. This indicates that the conditions for first-order constructive interference would not exist.

ADDITIONAL PRACTICE

1. Comet Hale-Bopp, which came close to Earth in 1997, has a complex chemical composition. To understand it, scientists analyzed radiation emitted from the comet's nucleus. Carbon atoms in the comet emitted radiation with a wavelength of 156.1 nm. Using a double-slit apparatus with a slit separation of 1.20×10^3 nm to measure these wavelengths, at what angle would a fifth-order maximum be observed?

2. A typical optic fiber has a thickness of only 6.00×10^3 nm. Consider a beam from a standard He-Ne laser that has a wavelength equal to 633 nm. Suppose this beam is incident upon two parallel slits that are separated by a distance equal to the width of a typical optic fiber. What is the angle at which the first dark fringe would be observed?

3. The smallest printed and bound book, which contains the children's story "Old King Cole," was published in 1985. The book's width is about 0.80 mm. Imagine a double-slit apparatus with a separation equal to the width of this book. What wavelength would produce a third-order minimum at an angle of 1.6°?

4. The water in Earth's atmosphere blocks most of the infrared waves coming from space. In order to observe light of this wavelength, the Kuiper Airborne Observatory has been developed. The observatory consists of an optical telescope mounted inside a modified C-141 aircraft. The plane flies at altitudes where the relative humidity is very low and where the incoming infrared radiation has not yet been significantly absorbed. Suppose a double-slit arrangement with a 15.0 μm slit separation is used to analyze infrared waves received by the telescope and that a second-order maximum is observed at 19.5°. Determine the wave's wavelength.

5. In 1995, Pan Xixing, of China, set a record for miniature writing by placing the text of a Chinese proverb on a human hair. If blue light with a wavelength of 443 nm passes through two slits separated by a distance equal to the width of an average character in that proverb, the fourth-order minimum would be observed at an angle of 2.27°. Determine the average width of each character.

6. A way to detect termites inside wooden structures has been developed at the University of Minnesota. A detector perceives the high-frequency sound waves produced by the termites' chewing. These waves are then converted into electromagnetic signals. Suppose these signals are at "long" radio wavelengths and that a giant double-slit apparatus has been built to observe interference of these radio waves. If the waves have a frequency of 60.0 kHz, what would be the required slit separation for observing the fourth-order maximum at 52.0°?

7. The Federal Communications Commission (FCC) assigns radio frequencies to broadcasters to prevent stations close to each other from transmitting at the same frequency. One reserved portion of the radio spectrum is just beyond commercial FM frequencies and is used by weather satellites. These broadcast at frequencies close to 137 MHz. Suppose radiation with this frequency is incident on a double-slit apparatus and that a second-order maximum is observed at 60.0°. What is the slit separation? What is the highest order maximum that can be observed for this radiation and with this apparatus?

Interference and Diffraction

Problem B

DIFFRACTION GRATINGS

PROBLEM

Two graduate students from the University of Nevada have developed Venetian blinds that open and close automatically by use of a solar sensor. Open blinds let in infrared radiation, which increases the temperature of the room. Consider the blinds as a type of diffraction grating with a line separation equal to 5.0 cm. Suppose the infrared waves pass through this grating so that the third-order maximum is observed at an angle of 0.69°. What is the wavelength of the infrared radiation?

SOLUTION

1. DEFINE **Given:** $\theta = 0.69°$

$d = 5.0 \text{ cm} = 5.0 \times 10^{-2} \text{ m}$

$m = 3$

Unknown: $\lambda = ?$

2. PLAN **Choose the equation(s) or situation:** Use the equation for a diffraction grating.

$$d\sin\theta = m\lambda$$

Rearrange the equation(s) to isolate the unknown(s):

$$\lambda = \frac{d\sin\theta}{m}$$

3. CALCULATE **Substitute the values into the equation(s) and solve:**

$$\lambda = \frac{(5.0 \times 10^{-2} \text{ m})(\sin 0.69°)}{3}$$

$$\lambda = \boxed{2.0 \times 10^{-4} \text{ m} = 0.20 \text{ mm}}$$

4. EVALUATE The electromagnetic radiation that is diffracted slightly by the Venetian blinds is in the infrared portion of the spectrum. To increase the angle at which the diffraction fringes appear, it would be necessary to narrow the spacing between the slats in the blinds.

ADDITIONAL PRACTICE

1. In 1996, a phone call was placed from the U.S. National Military Command Center to the U.S. Atlantic Command, 304 km away. The phone signal, however, traveled 120 750 km because it was transmitted via a Milstar communication satellite. The satellite uses superhigh-frequency electromagnetic radiation to ensure reliable communication that is secured against eavesdropping. Assume these waves are passed through a diffraction grating with 1.00×10^2 lines/m. The first-order maximum appears at an angle of 30.0°. Determine the wavelength and frequency of the electromagnetic waves.

2. Micropipette tubes with an outer diameter of 0.02 μm are used in research with living cells. Imagine a diffraction grating with a line separation of 0.020 μm. If this grating is used to analyze electromagnetic radiation, what wavelength would produce a third-order maximum at an angle of 12°?

3. Most stars are believed to be very nearly spherical, but R Cassiopeiae is much closer to having an oblong, or oval, shape. This remarkable fact was discovered by photographing its image in the orange region of the visible spectrum (at a wavelength of 714 nm). Suppose this radiation is passed through a diffraction grating so that it produces a third-order maximum at an angle of 12.0°. What is the line separation in the grating?

4. In 1995, a satellite called the *X Ray Timing Explorer* (XTE) was launched by NASA. The satellite can analyze X rays coming from hot matter surrounding massive and often compact objects, such as neutron stars and black holes. Because X rays have such short wavelengths, the only diffraction gratings with sufficiently small slit separations are crystal lattices. In crystals, the separation between atoms is small enough to diffract X rays. Suppose X rays with a wavelength of 40.0 nm are incident on a crystal lattice that has a "line" separation of 150.0 nm. At what angle will the second-order maximum occur?

5. Electromagnetic radiation comes to Earth from hydroxyl radicals in giant molecular clouds in space. The frequency of this radiation is 1612 MHz, which is close to the microwave frequencies used for aircraft communications. Filtering out this background noise is an important task. Suppose a technique similar to visible spectroscopy is performed with ambient microwave radiation using a microwave radio telescope. These waves could be directed toward a large diffraction grating so that their wavelengths and intensities could be analyzed. If the grating's line separation is 45.0 cm, at what angle would the first-order maximum occur for microwaves with a frequency of 1612 MHz?

6. A new astronomical facility is set for completion on Mount Wilson, in Georgia, by 1999. Named the Center for High Angular Resolution Astronomy (CHARA) Array, it comprises five telescopes whose individual images will be analyzed by computer. The resulting data will then be used to form a single high-resolution image of distant galaxies. The technique is similar to what is currently being done with radio telescopes. The CHARA Array will be used to observe radiation in the infrared portion of the spectrum with wavelengths as short as 2200 nm. Suppose 2200 nm light passes through a diffraction grating with 64×10^3 lines/m and produces a fringe at an angle of 34.0°. What order maximum will the fringe be?

7. A new technology to improve the image quality of large-screen televisions has been developed recently. The entire screen would consist of tiny cells, each equipped with a movable mirror. The mirrors would change positions relative to the incoming signal, providing a brighter image. The entire screen would look like a diffraction grating with 250 000 lines/m. Imagine shining red light with a wavelength of 750 nm onto this grating. What order maximum would be observed at an angle of 48.6°?

NAME ______________________ DATE ____________ CLASS ______________

Electric Forces and Fields

Problem A

COULOMB'S LAW

PROBLEM

Suppose you separate the electrons and protons in a gram of hydrogen and place the protons at Earth's North Pole and the electrons at Earth's South Pole. How much charge is at each pole if the magnitude of the electric force compressing Earth is 5.17×10^5 N? Earth's diameter is 1.27×10^7 m.

SOLUTION

1. DEFINE **Given:**

$F_{electric} = 5.17 \times 10^5$ N

$r = 1.27 \times 10^7$ m

$k_C = 8.99 \times 10^9$ N•m^2/C^2

Unknown: $q = ?$

2. PLAN **Choose the equation(s) or situation:** Rearrange the magnitude of the electric force using Coulomb's law.

$$q = \sqrt{\frac{F_{electric}r^2}{k_C}}$$

3. CALCULATE **Substitute the values into the equation(s) and solve:** .

$$q = \sqrt{\frac{(5.17 \times 10^5 \text{ N})(1.27 \times 10^7 \text{ m})^2}{8.99 \times 10^9 \text{ N•m}^2/\text{C}^2}}$$

$$q = \boxed{9.63 \times 10^4 \text{ C}}$$

4. EVALUATE The electrons and the protons have opposite signs, so the electric force between them is attractive. The large size of the force (equivalent to the weight of a 52 700 kg mass at Earth's surface) indicates how strong the attraction between opposite charges in atoms is.

ADDITIONAL PRACTICE

1. The safe limit for beryllium in air is 2.0×10^{-6} g/m^3, making beryllium one of the most toxic elements. The charge on all electrons in the Be contained in 1 m^3 of air at the safe level is about 0.085 C. Suppose this charge is placed 2.00 km from a second charge. Calculate the value of the second charge if the magnitude of the electric force between the two charges is 8.64×10^{-8} N.

2. Kalyan Ramji Sain, of India, had a mustache that measured 3.39 m from end to end in 1993. Suppose two charges, q and $3q$, are placed 3.39 m apart. If the magnitude of the electric force between the charges is 2.4×10^{-6} N, what is the value of q?

3. The remotest object visible to the unaided eye is the great galaxy Messier 31 in the constellation Andromeda. It is located 2.4×10^{22} m from Earth. (By comparison, the sun is only about 1.5×10^{11} m away.) Suppose two clouds containing equal numbers of electrons are separated by a distance of 2.4×10^{22} m. If the magnitude of the electric force between the clouds is 1.0 N, what is the charge of each cloud?

4. In 1990, a French team flew a kite that was 1034 m long. Imagine two charges, +2.0 nC and −2.8 nC, at opposite ends of the kite. Calculate the magnitude of the electric force between them. If the separation of charges is doubled, what absolute value of equal and opposite charges would exert the same electric force?

5. Betelgeuse, one of the brightest stars in the constellation of Orion, has a diameter of 7.0×10^{11} m (500 times the diameter of the sun). Consider two compact clouds with opposite charge equal to 1.0×10^{5} C. If these clouds are located 7.0×10^{11} m apart, what is the magnitude of the electric force of attraction between them?

6. An Italian monk named Giovanni Battista Orsenigo was also a dentist. From 1868 to 1903 he extracted exactly 2 000 744 teeth, which on average amounts to about 156 teeth per day. Consider a group of protons equal to the total number of teeth. If this group is divided in half, calculate the charge of each half. Also calculate the magnitude of the electric force that would result if the two groups of charges are placed 1.00 km apart.

7. The business district of London has about 4000 residents. However, every business day about 320 000 people are there. Consider a group of 4.00×10^{3} protons and a group of 3.20×10^{5} electrons that are 1.00 km apart. Calculate the magnitude of the electric force between them. Calculate the magnitude of the electric force if each group contains 3.20×10^{5} particles and if the separation distance remains the same.

8. In 1994, element 111 was discovered by an international team of physicists. Its provisional name was unununium (Latin for "one-one-one"). Find the distance between two equal and opposite charges, each having a magnitude equal to the charge of 111 protons, if the magnitude of the electric force between them is 2.0×10^{-28} N.

9. By 2005, the world's tallest building will be the International Finance Center in Taipei, Republic of China. Suppose a 1.00 C charge is placed at both the base and the top of the International Finance Center. If the magnitude of the electric force stretching the building is 4.48×10^{4} N, how tall is the International Finance Center?

10. A 44 000-piece jigsaw puzzle was assembled in France in 1992. Suppose the puzzle were square in shape, and that a 5.00 nC charge is placed at the upper right corner of the puzzle and a charge of −2.50 nC is placed at the lower left corner. If the magnitude of the electric force the two charges exert on each other were 1.18×10^{-11} N, what would be the distance between the two charges? What would be the length of the puzzle's sides?

NAME ____________________ DATE ____________ CLASS ____________

Electric Forces and Fields

Problem B

THE SUPERPOSITION PRINCIPLE

PROBLEM

The cinema screen installed at the Science Park, in Taejon, Korea, is 24.7 m high and 33.3 m wide. Consider the arrangement of charges shown below. If $q_1 = 2.00$ nC, $q_2 = -3.00$ nC, and $q_3 = 4.00$ nC, find the magnitude and direction of the resultant electric force on q_1.

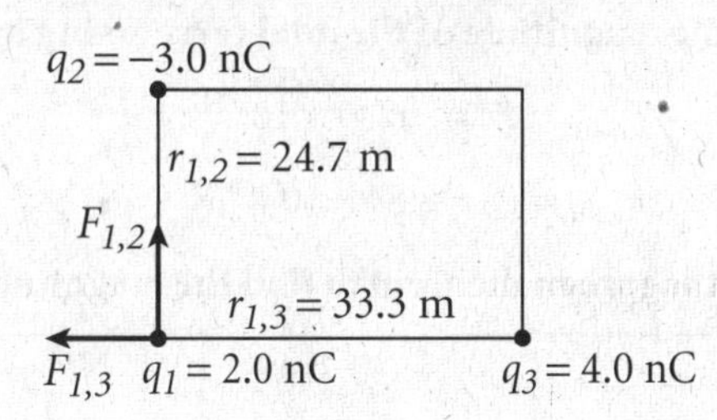

REASONING

According to the superposition principle, the resultant force on the charge q_1 is the vector sum of the forces exerted by q_2 and q_3 on q_1. First find the force exerted on q_1 by each charge, then use the Pythagorean theorem to find the magnitude of the resultant force on q_1. Take the ratio of the resultant y component to the resultant x component, and then take the arc tangent of that quantity to find the direction of the resultant force on q_1.

SOLUTION

Given:

$q_1 = 2.00 \text{ nC} = 2.00 \times 10^{-9} \text{ C}$

$q_2 = -3.00 \text{ nC} = -3.00 \times 10^{-9} \text{ C}$

$q_3 = 4.00 \text{ nC} = 4.00 \times 10^{-9} \text{ C}$

$r_{1,2} = 24.7 \text{ m}$

$r_{1,3} = 33.3 \text{ m}$

$k_C = 8.99 \times 10^9 \text{ N}\bullet\text{m}^2/\text{C}^2$

Unknown: $F_{1,tot} = ?$

Diagram:

$F_{1,2}$ $F_{1,3}$ q_1 → $F_{1,3}$ $F_{1,2}$ $F_{1,tot}$ q_1

1. Calculate the magnitude of the forces with Coulomb's law:

$$F_{2,1} = k_C\frac{q_2q_1}{(r_{2,1})^2} = \left(8.99 \times 10^9 \frac{\text{N}\bullet\text{m}^2}{\text{C}^2}\right)\left[\frac{(3.00 \times 10^{-9}\text{ C})(2.00 \times 10^{-9}\text{ C})}{(24.7\text{ m})^2}\right]$$

$$F_{2,1} = 8.84 \times 10^{-11}\text{ N}$$

$$F_{3,1} = k_C\frac{q_3q_1}{(r_{3,1})^2} = \left(8.99 \times 10^9 \frac{\text{N}\bullet\text{m}^2}{\text{C}^2}\right)\left[\frac{(4.00 \times 10^{-9}\text{ C})(2.00 \times 10^{-9}\text{ C})}{(33.3\text{ m})^2}\right]$$

$$F_{3,1} = 6.49 \times 10^{-11}\text{ N}$$

2. Determine the direction of the forces by analyzing the signs of the charges:

The force $\mathbf{F_{2,1}}$ is attractive because q_1 and q_2 have opposite signs. $\mathbf{F_{2,1}}$ is directed along the positive y-axis, so its sign is positive.

The force $\mathbf{F_{3,1}}$ is repulsive because q_1 and q_3 have the same sign. $\mathbf{F_{3,1}}$ is directed toward the negative x-axis, so its sign is negative.

3. Find the x and y components of each force:

For $\mathbf{F_{2,1}}$: $F_x = F_{3,1} = -6.49 \times 10^{-11}$ N; $F_y = 0$

For $\mathbf{F_{3,1}}$: $F_y = F_{2,1} = 8.84 \times 10^{-11}$ N; $F_x = 0$

4. Calculate the magnitude of the total force acting in both directions:

$$F_{x,tot} = F_x = -6.49 \times 10^{-11} \text{ N}$$

$$F_{y,tot} = F_y = 8.84 \times 10^{-11} \text{ N}$$

5. Use the Pythagorean theorem to find the magnitude of the resultant force:

$$F_{1,tot} = \sqrt{(F_{x,tot})^2 + (F_{y,tot})^2} = \sqrt{(-6.49 \times 10^{-11} \text{ N})^2 + (8.84 \times 10^{-11} \text{ N})^2}$$

$$F_{1,tot} = \boxed{1.10 \times 10^{-10} \text{ N}}$$

6. Use a suitable trigonometric function to find the direction of the resultant force:

In this case, you can use the inverse tangent function.

$$\tan\theta = \frac{F_{y,tot}}{F_{x,tot}} = \frac{(8.84 \times 10^{-11} \text{ N})}{(-6.49 \times 10^{-11} \text{ N})} = -1.36$$

$$\theta = \tan^{-1}(-1.36) = \boxed{-53.7°}$$

7. Evaluate your answer:

The resultant force makes an angle of 53.7° to the left and above the x-axis.

ADDITIONAL PRACTICE

1. In 1919 in Germany, a train of eight kites was flown 9740 m above the ground. This distance is 892 m higher than Mount Everest. Consider the arrangement of charges located at the various heights shown below. If $q_1 = 2.80$ mC, $q_2 = -6.40$ mC, and $q_3 = 48.0$ mC, find the magnitude and direction of the resultant electric force acting on q_1.

• $q_1 = 2.80$ mC

$r_{1,2} = 892$ m

• $q_2 = -6.40$ mC

$r_{1,3} = 9740$ m

• $q_3 = 48.0$ mC

2. In 1994, a group of British and Canadian athletes performed a rope slide off the top of Mount Gibraltar, in Canada. The speed of the sliders at times exceeded 160 km/h. The total length of the slide was 1747 m. Suppose several sliders are located on the rope as shown. Due to friction, they acquire the electric charges shown. Find the magnitude and direction of the resultant electric force acting on the athlete at the far right of the diagram.

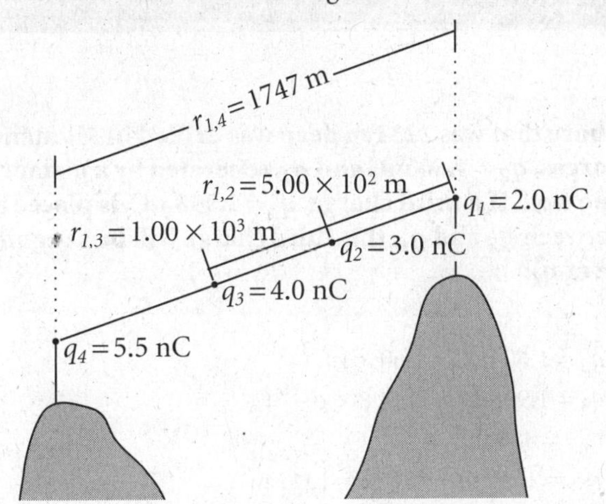

3. In 1913, a special postage stamp was issued in China. It was 248 mm long and 70.0 mm wide. Suppose equal charges of 1.0 nC are placed in the corners of this stamp. Find the magnitude and direction of the resultant electric force acting on the upper right corner (assume the widest part of the stamp is aligned with the *x*-axis).

4. In 1993, a chocolate chip cookie was baked in Arcadia, California. It contained about three million chips and was 10.7 m long and 8.7 m wide. Suppose four charges are placed in the corners of that cookie as follows: $q_1 = -12.0$ nC at the lower left corner, $q_2 = 5.6$ nC at the upper left corner, $q_3 = 2.8$ nC at the upper right corner, and $q_4 = 8.4$ nC at the lower right corner. Find the magnitude and direction of the resultant electric force acting on q_1.

5. In 1988, a giant firework was exploded at the Lake Toya festival, in Japan. The shell had a mass of about 700 kg and produced a fireball 1.2 km in diameter. Consider a circle with this diameter. Suppose four charges are placed on the circle's perimeter so that the lines between them form a square with sides parallel to the *x*- or *y*-axes. The charges have the following strengths and locations: $q_1 = 16.0$ mC at the upper left "corner," $q_2 = 2.4$ mC at the upper right corner, $q_3 = -3.2$ mC at the lower right corner, and $q_4 = -4.0$ mC at the lower left corner. Find the magnitude and direction of the resultant electric force acting on q_1. (Hint: Find the distances between the charges first.)

6. American athlete Jesse Castenada walked 228.930 km in 24 h in 1976, setting a new record. Consider an equilateral triangle with a perimeter equal to the distance Castenada walked. Suppose the charges are placed at the following vertices of the triangle: $q_1 = 8.8$ nC at the bottom left vertex, $q_2 = -2.4$ nC at the bottom right vertex, and $q_3 = 4.0$ nC at the top vertex. Find the magnitude and direction of the resultant electric force acting on q_1.

NAME ______________________ DATE ____________ CLASS ______________

Electric Forces and Fields

Problem C

EQUILIBRIUM

PROBLEM

In 1955, a water bore that was 2231 m deep was drilled in Montana. Consider two charges, q_2 = 1.60 mC and q_1, separated by a distance equal to the depth of the well. If a third charge, q_3 = 1.998 μC, is placed 888 m from q_2 and is between q_2 and q_1, this third charge will be in equilibrium. What is the value of q_1?

SOLUTION

1. DEFINE

Given:

$q_2 = 1.60 \text{ mC} = 1.60 \times 10^{-3} \text{ C}$

$q_3 = 1.998\ \mu\text{C} = 1.998 \times 10^{-6} \text{ C}$

$r_{3,2} = 888 \text{ m}$

$r_{3,1} = 2231 \text{ m} - 888 \text{ m} = 1342 \text{ m}$

$r_{2,1} = 2231 \text{ m}$

$k_C = 8.99 \times 10^9 \text{ N}\bullet\text{m}^2/\text{C}^2$

Unknown: $q_1 = ?$

Diagram:

$r_{2,1} = 2231$ m

$r_{3,1} = 1342$ m — $r_{3,2} = 888$ m

q_1 — $q_3 = 1.998\ \mu$C — $q_2 = 1.60$ mC

2. PLAN

Choose the equation(s) or situation: The force exerted on q_3 by q_2 will be opposite the force exerted on q_3 by q_1. The resultant force on q_3 must be zero in order for the charge to be in equilibrium. This indicates that $F_{3,1}$ and $F_{3,2}$ must be equal to each other.

$$F_{3,1} = k_C\left(\frac{q_3 q_1}{(r_{3,1})^2}\right) \text{ and } F_{3,2} = k_C\left(\frac{q_3 q_2}{(r_{3,2})^2}\right)$$

$$F_{3,1} = F_{3,2}$$

$$k_C\left(\frac{q_3 q_1}{(r_{3,1})^2}\right) = k_C\left(\frac{q_3 q_2}{(r_{3,2})^2}\right)$$

Rearrange the equation(s) to isolate the unknown(s): q_3 and k_C cancel.

$$q_1 = q_2\left(\frac{r_{3,1}}{r_{3,2}}\right)^2$$

3. CALCULATE

Substitute the values into the equation(s) and solve:

$$q_1 = (1.60 \times 10^{-3} \text{ C})\left(\frac{1342 \text{ m}}{888 \text{ m}}\right)^2 = 3.65 \times 10^{-3} \text{ C}$$

$$q_1 = \boxed{3.65 \text{ mC}}$$

4. EVALUATE

Because q_1 is a little more than twice as large as q_2, the third charge (q_3) must be farther from q_1 for the forces on q_3 to balance.

ADDITIONAL PRACTICE

1. Hans Langseth's beard measured 5.33 m in 1927. Consider two charges, $q_1 = 2.50$ nC and an unspecified charge, q_2, are separated 5.33m. A third charge of 1.0 nC is placed 1.90 m away from q_1. If the net electric force on this third charge is zero, what is q_2?

2. The extinct volcano Olympus Mons, on Mars, is the largest mountain in the solar system. It is 6.00×10^2 km across and 24 km high. Suppose a charge of 75 mC is placed 6.0×10^2 km from a unspecified charge. If a third charge of 0.10 mC is placed 24 km from the first charge and the net electric force on this third charge is zero, how large is the unspecified charge?

3. Earth's mass is about 6.0×10^{24} kg, while the moon's mass is 7.3×10^{22} kg. What equal charges must be placed on Earth and the moon to make the net force between them zero?

4. In 1974, an emerald with a mass of 17.23 kg was found in Brazil. Suppose you want to hang this emerald on a string that is 80.0 cm long and has a breaking strength of 167.6 N. To hang the jewel safely, you remove a certain charge from the emerald and place it at the pivot point of the string. What is the minimum possible value of this charge?

5. Little Pumpkin, a miniature horse owned by J. C. Williams, Jr., of South Carolina, had a mass of about 9.00 kg. Consider Little Pumpkin on a twin-pan balance. If the mass on the other pan is 8.00 kg and r equals 1.00 m, what equal and opposite charges must be placed as shown in the diagram below to maintain equilibrium?

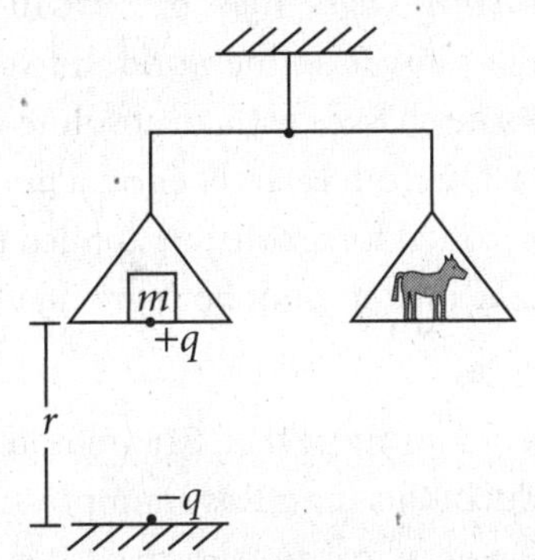

6. The largest bell that is in use today is in Mandalay, Myanmar, formerly called Burma. Its mass is about 92×10^3 kg. Suppose the bell is supported in equilibrium as shown in the figure below. Calculate the value for the charge q.

$l_1 = 1.0$ m

$+q$

$l_2 = 8.0$ m

$r = 2.5$ m

$m = 92 \times 10^3$ kg

$-q$

7. In more than 30 years, Albert Klein, of California, drove 2.5×10^6 km in one automobile. Consider two charges, $q_1 = 2.0$ C and $q_2 = 6.0$ C, separated by Klein's total driving distance. A third charge, $q_3 = 4.0$ C, is placed on the line connecting q_1 *and* q_2. How far from q_1 should q_3 be placed for q_3 to be in equilibrium?

8. A 55 μC charge and a 137 μC charge are separated by 87 m. Where must a 14 μC charge be placed between these other two charges in order for the net electric force on it to be zero?

9. In 1992, a Singapore company made a rope that is 56 cm in diameter and has an estimated breaking strength of 1.00×10^8 N. Suppose this rope is used to connect two strongly charged asteroids in space. If the charges of the asteroids are $q_1 = 1.80 \times 10^4$ C and $q_2 = 6.25 \times 10^4$ C, what is the minimum length that the rope can have and still remain intact? Neglect the effects of gravity.

10. The CN Tower, in Toronto, Canada, is 553 m tall. Suppose two balls, each with a mass of 5.00 kg and a charge of 40.0 mC, are placed at the top and bottom of the tower, respectively. The ball at the top is then dropped. At what height is the acceleration on the ball zero?

11. *Mycoplasma* is the smallest living organism known. Its mass has an estimated value of 1.0×10^{-16} g. Suppose two specimens of this organism are placed 1.0 m apart and one electron is placed on each. If the medium through which the *Mycoplasma* move exerts a resistive force on the organisms, how large must that force be to balance the force of electrostatic repulsion?

12. The parasitic wasp *Carapractus cinctus* has a mass of 5.0×10^{-6} kg, which makes it one of the smallest insects in the world. If two such wasps are given equal and opposite charges with an absolute value of 2.0×10^{-15} C and are placed 1.00 m from each other on a horizontal smooth surface, what extra horizontal force must be applied to each wasp to keep it from sliding? Take into account both gravitational and electric forces between the wasps.

13. In 1995, a single diamond was sold for more than $16 million. It was not the largest diamond in the world, but its mass was an impressive 20.0 g. Consider such a diamond resting on a horizontal surface. It is known that if the diamond is given a charge of 2.0 μC and a charge of at least −8.0 μC is placed on that surface at a distance of 1.7 m from it, then the diamond will barely keep from sliding. Calculate the coefficient of static friction between the diamond and the surface.

Electric Forces and Fields

Problem D

ELECTRIC FIELD STRENGTH

PROBLEM

The Seto-Ohashi bridge, linking the two Japanese islands of Honshu and Shikoku, is the longest "rail and road" bridge, with an overall length of 12.3 km. Suppose two equal charges are placed at the opposite ends of the bridge. If the resultant electric field strength due to these charges at the point exactly 12.3 km above one of the bridge's ends is 3.99×10^{-2} N/C and is directed at 75.3° above the positive *x*-axis, what is the magnitude of each charge?

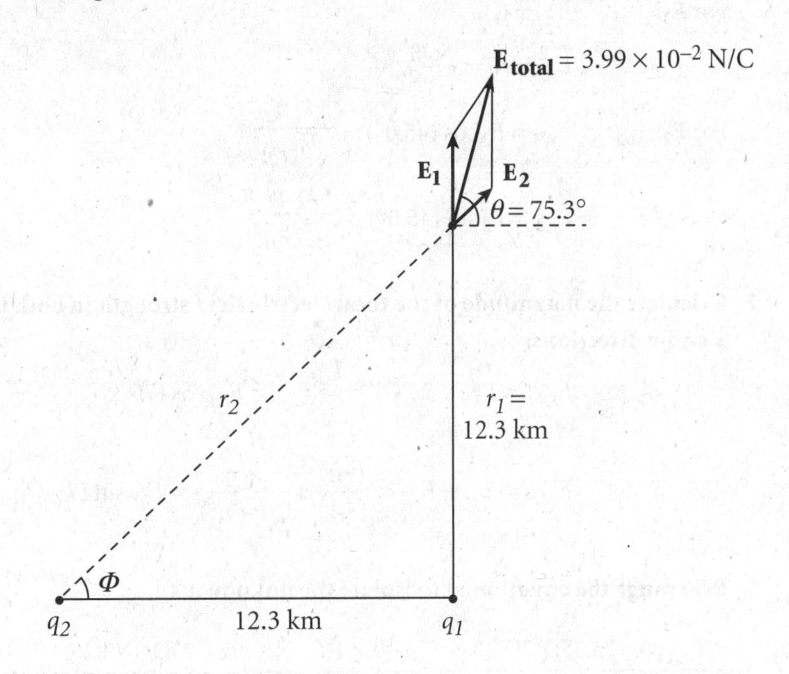

REASONING

According to the superposition principle, the resultant electric field strength at the point above the bridge is the vector sum of the electric field strengths produced by q_1 and q_2. First, find the components of the electric field strengths produced by each charge, then combine components in the *x* and *y* directions to find the electric field strength components of the resultant vector. Equate this to the components in the *x* and *y* directions of the electric field vector. Finally, rearrange the equation to solve for the charge.

SOLUTION

Given:

$E_{tot} = 3.99 \times 10^{-4}$ N/C

$\theta = 75.3°$

$r_1 = 12.3 \text{ km} = 1.23 \times 10^4$ m

$k_C = 8.99 \times 10^9 \text{ N}\bullet\text{m}^2/\text{C}^2$

Unknown: $q_1 = ?$ $q_2 = ?$

The distance r_2 must be calculated from the information in the diagram. Because r_2 forms the hypotenuse of a right triangle whose sides equal r_1, it follows that

$$r_2 = \sqrt{(r_1)^2 + (r_1)^2} = \sqrt{2(r_1)^2} = 1.74 \times 10^4 \text{ m}$$

The angle that r_2 makes with the coordinate system equals the inverse tangent of the ratio of the vertical to the horizontal components. Because these components are equal,

$$\tan \phi = 1.00, \text{ or } \phi = 45.0^\circ$$

1. **Find the *x* and *y* components of each electric field strength vector:**
   At this point, the direction of each component must be taken into account.

For $\mathbf{E_1}$:
$$E_{x,1} = 0$$
$$E_{y,1} = E_1 = \frac{k_C q_1}{(r_1)^2}$$

For $\mathbf{E_2}$:
$$E_{x,2} = E_2 \cos(45.0^\circ) = \frac{k_C q_2}{\sqrt{2}(r_2)^2}$$
$$E_{y,2} = E_2 \sin(45.0^\circ) = \frac{k_C q_2}{\sqrt{2}(r_2)^2}$$

2. **Calculate the magnitude of the total electric field strength in both the *x* and *y* directions:**

$$E_{x,tot} = E_{x,1} + E_{x,2} = \frac{k_C q_2}{\sqrt{2}(r_2)^2} = E_{tot} \cos(75.3^\circ)$$

$$E_{y,tot} = E_{y,1} + E_{y,2} = \frac{k_C q_1}{(r_1)^2} + \frac{k_C q_2}{\sqrt{2}(r_2)^2} = E_{tot} \sin(75.3^\circ)$$

5. **Rearrange the equation(s) to isolate the unknown(s):**

$$q_2 = \frac{E_{tot} \cos(75.3^\circ)\sqrt{2}(r_2)^2}{k_C} = \frac{(3.99 \times 10^{-2} \text{ N/C})\cos(75.3^\circ)\sqrt{2}(1.74 \times 10^4 \text{ m})^2}{8.99 \times 10^9 \text{ N}\bullet\text{m}^2/\text{C}^2}$$

$$q_2 = q_1 = \boxed{4.82 \times 10^{-4} \text{ C}}$$

NAME ______________________ DATE ____________ CLASS ____________

ADDITIONAL PRACTICE

1. The world's largest tires have a mass of almost 6000 kg and a diameter of 3.72 m each. Consider an equilateral triangle with sides that are 3.72 m long each. If equal positive charges are placed at the points on either end of the triangle's base, what is the direction of the resultant electric field strength vector at the top vertex? If the magnitude of the electric field strength at the top vertex equals 0.145 N/C, what are the two quantities of charge at the base of the triangle?

2. The largest fountain is found at Fountain Hills, Arizona. Under ideal conditions, the 8000 kg column of water can reach as high as 190 m. Suppose a 12 nC charge is placed on the ground and another charge of unknown quantity is located 190 m above the first charge. At a point on the ground 120 m from the first charge, the horizontal component of the resultant electric field strength is found to be $E_x = 1.60 \times 10^{-2}$ N/C. Using this information, calculate the unknown quantity of charge.

3. Pontiac Silverdome Stadium, in Detroit, Michigan, is the largest air-supported building in the world. Suppose a charge of 18.0 μC is placed at one end of the stadium and a charge of −12.0 μC is placed at the other end of the stadium. If the electric field halfway between the charges is 22.3 N/C, directed toward the −12.0 μC charge, what is the length of the stadium?

4. In 1897, a Ferris wheel with a diameter of 86.5 m was built in London. The wheel held 10 first-class and 30 second-class cabins, and each cabin was capable of carrying 30 people. Consider two cabins positioned exactly opposite each other. Suppose one cabin has an unbalanced charge of 4.8 nC and the other cabin has a charge of 16 nC. At what distance from the 4.8 nC charge along the diameter of the wheel would the strength of the resultant electric field be zero?

5. Suppose three charges of 3.6 μC each are placed at three corners of the Imperial Palace in Beijing, China, which has a length of 960 m and a width of 750 m. What is the strength of the electric field at the fourth corner?

6. The world's largest windows, which are in the Palace of Industry and Technology in Paris, France, have a maximum width of 218 m and a maximum height of 50.0 m. Consider a rectangle with these dimensions. If charges are placed at its corners, as shown in the figure below, what is the electric field strength at the center of the rectangle? The value of q is 6.4 nC.

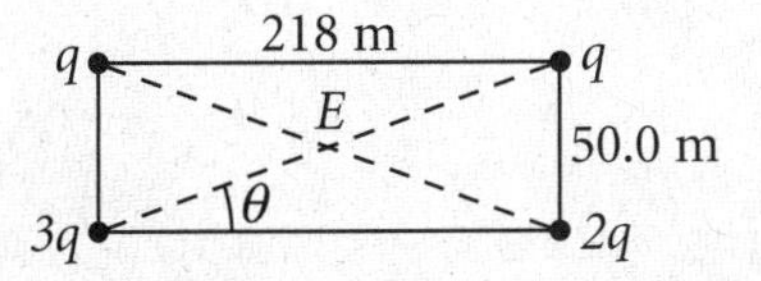

NAME ______________________ DATE ______________ CLASS ______________

Electrical Energy and Current

Problem A

POTENTIAL ENERGY AND POTENTIAL DIFFERENCE

PROBLEM

As a charge moves 39.0 cm in the direction of a uniform electric field, its potential energy changes by -1.86×10^{-3} J. The magnitude of the electric field is 177 N/C. Find the potential difference between the starting position and final position of the charge. What is the charge on the moving particle?

SOLUTION

Given: $\Delta PE_{electric} = -1.86 \times 10^{-3}$ J $\quad d = 0.390$ m

$E = 177$ N/C

Unknown: $\Delta V = ?$ $\quad q = ?$

The potential difference is the magnitude of E times the displacement.

$$\Delta V = -Ed = -(177 \text{ N/C})(0.390 \text{ m})$$

$$\boxed{\Delta V = -69.0 \text{ V}}$$

Use the equation for the change in electrical potential energy.

$$\Delta PE_{electric} = -qEd$$

Rearrange to solve for q, and insert values.

$$q = -\frac{\Delta PE_{electric}}{Ed} = -\frac{(-1.86 \times 10^{-3} \text{ J})}{(177 \text{ N/C})(0.390 \text{ m})}$$

$$\boxed{q = 2.69 \times 10^{-5} \text{ C}}$$

ADDITIONAL PRACTICE

1. What is the potential difference between a point 0.48 mm from a charge of 2.9 nC and a point at infinity?

2. The potential energy of an electron ($q = -1.6 \times 10^{-19}$ C) increases by 3.3×10^{-15} J when it moves 3.5 cm parallel to a uniform electric field. What is the magnitude of the electric field through which the electron passes?

3. A charged particle gains 3.1×10^{-12} J of potential energy when it moves 4.7 cm through a uniform electric field. The potential difference across this distance is –73 V.

a. What is the magnitude of the electric field?

b. What is the magnitude of the charge on the particle?

4. A charged particle moves through a distance of 9.35 m parallel to a uniform electric field. The electrical potential energy of the particle increases by 3.17×10^{-10} J as it moves. The electric field has a magnitude of 1.25×10^{5} N/C.

a. What is the charge on the particle?

b. What is the potential difference between the starting and final positions?

5. There is an electric field close to the surface of Earth. This field points toward the surface and has a magnitude of about 1.5×10^{2} N/C. A charge moves perpendicularly toward the surface of Earth through a distance of 439 m, the height of the Sears Tower in Chicago, Illinois. During this trip, the electric potential energy of the charge decreases by 3.7×10^{-8} J.

a. What is the charge on the moving particle?

b. What is the potential difference between the top of the Sears Tower and the ground?

c. What is the electric potential of the charge at its final position?

6. Two parallel plates are separated by a distance of 0.077 cm. Between the plates is a uniform electric field having a magnitude of 6.5×10^{2} N/C. What is the magnitude of the potential difference between the plates?

7. A proton ($q = 1.6 \times 10^{-19}$ C) moves 3.75 m in the direction of a uniform electric field having a field strength of 383 N/C.

a. What is the magnitude of the potential difference between the initial and final positions of the proton?

b. What is the change in the electrical potential energy of the proton?

8. A negative ion ($q = -4.8 \times 10^{-19}$ C) moves 0.63 cm through a uniform electric field in a direction opposite to the direction of the field. The magnitude of the electric field is 279 volts per meter.

a. What is the change in the electrical potential energy of the ion?

b. What is the electric potential at the final position of the ion?

NAME ______________________ DATE ______________ CLASS ______________

Electrical Energy and Current

Problem B

CAPACITANCE

PROBLEM

Consider a parallel-plate capacitor the size of Nauru, an island with an area of just 21 km^2. If 110 V is applied across the capacitor to store 55 J of electrical potential energy, what is the capacitance of this capacitor? If the area of its plates is the same as the area of Nauru, what is the plate separation?

SOLUTION

Given:

$A = 21\ \text{km}^2 = 21 \times 10^6\ \text{m}^2$

$PE_{electric} = 55\ \text{J}$

$\Delta V = 110\ \text{V}$

$\varepsilon_0 = 8.85 \times 10^{-12}\ \text{C}^2/\text{N}\bullet\text{m}^2$

Unknown: $C = ?$ $\quad d = ?$

To determine capacitance and the plate separation, use the definition for capacitance energy and the equation for a parallel-plate capacitor.

$$PE_{electric} = \frac{1}{2}C(\Delta V)^2$$

$$C = \frac{2\,PE_{electric}}{(\Delta V)^2} = \frac{2(55\ \text{J})}{(110\ \text{V})^2} = \boxed{9.1 \times 10^{-3}\ \text{F}}$$

$$C = \varepsilon_0 \frac{A}{d}$$

$$d = \varepsilon_0 \frac{A}{C}$$

$$d = \left(8.85 \times 10^{-12}\ \frac{\text{C}^2}{\text{N}\bullet\text{m}^2}\right)\frac{(21 \times 10^6\ \text{m}^2)}{(9.1 \times 10^{-3}\ \text{F})}$$

$$d = \boxed{2.0 \times 10^{-2}\ \text{m} = 2.0\ \text{cm}}$$

ADDITIONAL PRACTICE

1. To improve the short-range acceleration of an electric car, a capacitor may be used. Charge is stored on the capacitor's surface between a porous composite electrode and electrolytic fluid. Such a capacitor can provide a potential difference of nearly 3.00×10^2 V. If the energy stored in this capacitor is 17.1 kJ, what is the capacitance?

2. A huge capacitor bank powers the large Nova laser at Lawrence Livermore National Lab, in California. Each capacitor can store 1450 J of electrical potential energy when a potential difference of 1.0×10^4 V is applied across its plates. What is the capacitance of this capacitor?

3. Air becomes a conductor if the electric field across it exceeds 3.0×10^6 V/m. What is the maximum charge that can be accumulated on an air-filled capacitor with a 0.2 mm plate separation and a plate area equal to the area of the world's largest painting (6.7×10^3 m^2)?

4. The world's largest hamburger, which was made in Wyoming, had a radius of about 3.1 m. Suppose you build an air-filled parallel-plate capacitor with an area equal to that of the hamburger and a plate separation of 1.0 mm. The largest electric field that air can sustain before its insulating properties break down and it begins conducting electricity is 3.0 MV/m. What is the maximum charge that you will be able to store in the capacitor?

5. In 1987, an ultraviolet flash of light was produced at the Los Alamos National Lab. For 1.0 ps, the power of the flash was 5.0×10^{15} W. If all of this energy was provided by a battery of charged capacitors with a total capacitance of 0.22 F, what was the potential difference across these capacitors?

6. The International Exposition Center in Cleveland, Ohio, covers 2.32×10^5 m^2. If a capacitor with this same area and a plate separation of 1.5 cm is charged to 0.64 mC, what is the energy stored in the capacitor?

7. In 1990, a pizza with a radius of 18.0 m was made in South Africa. Suppose you make an air-filled capacitor with parallel plates whose area is equal to that of the pizza. If a potential difference of 575 V is applied across this capacitor, it will store just 3.31 J of electrical potential energy. What are the capacitance and plate separation of this capacitor?

8. When the keys on a computer keyboard are pressed, the plate separation of small parallel-plate capacitors mounted under the keys changes from about 5.00 mm to 0.30 mm. This causes a change in capacitance, which triggers the electronic circuitry. If the area of the plates is 1.20 cm^2, find the change in the capacitance. Is it positive or negative?

9. Tristan da Cunha, a remote island inhabited by a few hundred people, has an area of 98 km^2. Suppose a 0.20 F parallel-plate capacitor with a plate area equal to 98 km^2 is built. What is the plate separation?

10. The largest apple pie ever made was rectangular in shape and had an area of 7.0 m $\times$ 12.0 m. Consider a parallel-plate capacitor with this area. Assume that the capacitor is filled with air and the distance between the plates is 1.0 mm.

- **a.** What is the capacitance of the capacitor?
- **b.** What potential difference would have to be placed across the capacitor for it to store 1.0 J of electrical potential energy?

11. The largest lasagna in the world was made in California in 1993. It had an area of approximately 44 m^2. Imagine a parallel-plate capacitor with this area that is filled with air. If a potential difference of 30.0 V is placed across the capacitor, the stored charge is 2.5 μC.

a. Calculate the capacitance of the capacitor.

b. Calculate the distance that the plates are separated.

c. Calculate the electrical potential energy that is stored in the capacitor.

Electrical Energy and Current

Problem C

CURRENT

PROBLEM

Amtrak introduced an electric train in 2000 that runs between New York and Boston. With a travel time of 3.00 h, the train is not superfast, but it is comfortable, very safe, and environmentally friendly. Find the charge passing through the engine during the trip if the feeding current is 1.80×10^2 A.

SOLUTION

Given: $\Delta t = 3.00 \text{ h} = 1.08 \times 10^4 \text{ s}$

$I = 1.80 \times 10^2 \text{ A}$

Unknown: $\Delta Q = ?$

Use the equation for electric current.

$$I = \frac{\Delta Q}{\Delta t}$$

$$\Delta Q = I\Delta t = (1.80 \times 10^2 \text{ A})(1.08 \times 10^4 \text{ s}) = \boxed{1.94 \times 10^6 \text{ C}}$$

ADDITIONAL PRACTICE

1. An 8.00×10^6 kg electromagnet built in Switzerland draws a current of 3.00×10^2 A. How much charge passes through the magnet in 2.4 min?
2. On July 16, 2186, a total solar eclipse lasting 7 min, 29 s, will be observed over the mid-Atlantic Ocean. During the eclipse, an observer may need a flashlight to see. If the flashlight draws a current of 0.22 A, how much charge will pass through the light bulb during the eclipse?
3. The National Institute of Standards and Technology has built a clock that is off by only 3.3 μs a year. Consider a current of 0.88 A in a wire. How many electrons pass through a cross-section of the wire in 3.3×10^{-6} s?
4. In England, a miniature electric bicycle has been constructed. The Ni-Cd batteries help the rider pedal for up to 3.00 h. If a charge of 1.51×10^4 C passes through the motor during those 3.00 h, what is the current?
5. Because high-temperature superconducting cables cannot sustain high currents, their applicability is limited. However, a prototype of a high-temperature superconductor that can transfer a charge of 1.8×10^5 C in 6.0 min has been constructed. To what current does this correspond?
6. In 1994, a prototype of what was claimed to be the first "practical" electric car was introduced. Recharging the batteries is not very practical, though. Calculate the average time needed to recharge the car's batteries if a 13.6 A current carries 4.40×10^5 C to the batteries.

NAME ______________________ DATE ______________ CLASS ______________

Electrical Energy and Current

Problem D

RESISTANCE

PROBLEM

A medical belt pack with a portable laser for in-the-field medical purposes has been constructed. The laser draws a current of 2.5 A and the circuitry resistance is 0.6 Ω. What is the potential difference across the laser?

SOLUTION

Given: $I = 2.5\ \text{A}$ $R = 0.6\ \Omega$

Unknown: $\Delta V = ?$

Use the definition of resistance.

$$\Delta V = IR = (2.5\ \text{A})(0.6\ \Omega) = \boxed{1.5\ \text{V}}$$

ADDITIONAL PRACTICE

1. Electric eels, found in South America, can provide a potential difference of 440 V that draws a current of 0.80 A through the eel's prey. Calculate the resistance of the circuit (the eel and prey).
2. It is claimed that a certain camcorder battery can provide a potential difference of 9.60 V and a current of 1.50 A. What is the resistance through which the battery must be discharged?
3. A prototype electric car is powered by a 312 V battery pack. What is the resistance of the motor circuit when 2.8×10^5 C passes through the circuit in 1.00 h?
4. In 1992, engineers built a 2.5 mm long electric motor that can be driven by a very low emf. What is the potential difference if it draws a 3.8 A current through a 0.64 Ω resistor?
5. A team from Texas A&M University has built an electric sports car with a maximum motor current of 2.4×10^3 A. Determine the potential difference that provides this current if the circuit resistance is 0.30 Ω.
6. Stanford University scientists have constructed the Orbiting Picosatellite Automated Launcher (OPAL). OPAL can launch disposable "picosatellites" the size of hockey pucks. Each picosatellite will be powered by a 3.0 V battery for about an hour. If the satellite's circuitry were to have a resistance of 16 Ω, what current would be drawn by the satellite?
7. For years, California has been striving for all zero-emission vehicles on its roads. In 1995, a street bus with a range of 120 km was built. This bus is powered by batteries delivering 6.00×10^2 V. If the circuit resistance is 4.4 Ω, what is the current in the bus's circuit?

Electrical Energy and Current

Problem E

ELECTRIC POWER

PROBLEM

In 1994, a group of students at Lawrence Technological University, in Southfield, Michigan, built a car that combines a conventional diesel engine and an electric direct-current motor. The power delivered by the motor is 32 kW. If the resistance of the car's circuitry is 8.0 Ω, find the current drawn by the motor.

SOLUTION

Given: $P = 32 \text{ kW} = 3.2 \times 10^4 \text{ W}$

$R = 8.0\ \Omega$

Unknown: $I = ?$

Because power and resistance are known, use the second form of the power equation to solve for I.

$$P = I^2R$$

$$I = \sqrt{\frac{P}{R}} = \sqrt{\frac{(3.2 \times 10^4 \text{ W})}{(8.0\ \Omega)}} = \boxed{63 \text{ A}}$$

ADDITIONAL PRACTICE

1. A flying source of light is being developed that will consist of a metal-halide lamp lifted by a helium-filled balloon. The maximum power rating for the lamp available for the device is 12 kW. If the lamp's resistance is $2.5 \times 10^2\ \Omega$, what is the current in the lamp?
2. The first American hybrid electric bus operated in New York in 1905. The gasoline-fueled generator delivered 33.6 kW to power the bus. If the generator supplied an emf of 4.40×10^2 V, how large was the current?
3. A compact generator has been designed that can jump-start a car, though the generator's mass is only about 10 kg. Find the maximum current that the generator can provide at 12.0 V if its maximum power is 850 W.
4. Fuel cells combine gaseous hydrogen and oxygen to effectively and cleanly produce energy. Recently, German engineers produced a fuel cell that can generate 4.2×10^{10} J of electricity in 1.1×10^3 h. What potential difference would this fuel cell place across a 40.0 Ω resistor?
5. Omega, a laser built at the University of Rochester, New York, generated 6.0×10^{13} W for 1.0 ns in 1995. If this power is provided by 8.0×10^6 V placed across the circuit, what is the circuit's resistance?
6. In 1995, Los Alamos National Lab developed a model electric power plant that used geothermal energy. Find the plant's projected power output if the plant produces a current of 6.40×10^3 A at 4.70×10^3 V.

NAME ______________________ DATE ____________ CLASS ______________

Electrical Energy and Current

Problem F

ELECTRIC POTENTIAL

PROBLEM

The longest railway platform is in Kharagpur, India. Suppose you place two charges, -8.1×10^{-8} C and -2.4×10^{-7} C, at opposite ends of this platform. The electric potential at a point 6.60×10^{2} m from the greater charge is -7.5 V. What is the distance between the charges?

SOLUTION

Given:

$q_1 = -8.1 \times 10^{-8}\ \text{C}$

$q_2 = -2.4 \times 10^{-7}\ \text{C}$

$r_2 = 6.60 \times 10^{2}\ \text{m}$

$V = -7.5\ \text{V}$

$k_C = 8.99 \times 10^{9}\ \text{N}\bullet\text{m}^2/\text{C}^2$

Unknown: $d = ?$

Use the equation for the electric potential near a point charge.

$$r_1 = d - r_2$$

$$V = k_C\frac{q}{r} = k_C\left(\frac{q_1}{r_1} + \frac{q_2}{r_2}\right) = k_C\left(\frac{q_1}{d - r_2} + \frac{q_2}{r_2}\right)$$

$$\frac{V}{k_C} - \frac{q_2}{r_2} = \frac{q_1}{d - r_2}$$

$$d = \frac{q_1}{\left(\frac{V}{k_c} - \frac{q_2}{r_2}\right)} + r_2$$

$$d = \frac{-8.1 \times 10^{-8}\ \text{C}}{\left[\frac{-7.5\ \text{V}}{(8.99 \times 10^{9}\ \text{N}\bullet\text{m}^2/\text{C}^2)} - \frac{(-2.4 \times 10^{-8}\ \text{C})}{660\ \text{m}}\right]} + 660\ \text{m}$$

$$d = \frac{-8.1 \times 10^{-8}\ \text{C}}{[-8.3 \times 10^{-10}\ \text{C/m} + 3.6 \times 10^{-10}\ \text{C/m}]} + 660\ \text{m}$$

$$d = \frac{-8.1 \times 10^{-8}\ \text{C}}{-4.7 \times 10^{-10}\ \text{C/m}} + 660\ \text{m} = \boxed{830\ \text{m}}$$

ADDITIONAL PRACTICE

1. The Sydney Harbour Bridge, in Australia, is the world's widest long-span bridge. Suppose two charges, -12.0 nC and -68.0 nC, are separated by the width of the bridge. The electric potential along a line between the charges at a distance of 16.0 m from the -12.0 nC charge is -25.3 V. How far apart are the charges?

2. The Hughes H4 Hercules, nicknamed the *Spruce Goose,* has a wingspan of 97.5 m, which is greater than the wingspan of any other plane. Suppose two charges, 18.0 nC and 92.0 nC, are placed at the tips of the wings. If an electric potential of 53.3 V is measured at a certain point along the wings, how far is that point from the 92.0 nC charge?

3. A Canadian company has developed a scanning device that can detect drugs, explosives, and even cash that are being smuggled across state borders. The scanner uses accelerated protons to generate gamma rays, which can easily penetrate through most substances that are less than a few centimeters thick. At the heart of the scanner is a compact power supply that can produce an electric potential as large as 1.0×10^6 V. Find the value of a point charge q that would create an electric potential of 1.0×10^6 V at a distance of 12 cm.

4. Gravitational potential and electric potential are described by similar mathematical equations. Suppose Earth has a uniform distribution of positive charge on its surface. A test particle with a mass of 1.0 kg and a charge of 1.0 C is placed at some distance from Earth. What must the total charge on Earth's surface be for the test particle to experience equal gravitational and electric potentials? (Hint: Obtain the equation for gravitational potential by comparing the gravitational force equation from Section 7I to Coulomb's law for electrostatic force. Assume that the entire charge on Earth's surface can be treated as a point charge at Earth's center.)

5. Overall, the matter in the sun is electrically neutral. However, the temperatures within the sun are too high for electrons to remain with positively charged nuclei for more than a fraction of a second. Suppose the positive and negative charges in the sun could be separated into two clouds with a separation of 1.5×10^8 km, which is equal to the average distance between Earth and the sun.
   - **a.** Calculate the charge in each cloud. Assume that the sun, which has a mass of 1.97×10^{30} kg, consists entirely of hydrogen. The mass of one hydrogen atom is 1.67×10^{-27} kg.
   - **b.** Calculate the electric potential for the two clouds at a distance of 1.1×10^8 km (the distance between the sun and Venus) from the proton cloud.

6. The freight station at Hong Kong's waterfront is the largest multilevel industrial building in the world. The station is 292 m long and 276 m wide, and inside there are more than 26 km of roadways. Consider a rectangle that is 292 m long and 276 m wide. If charges $+1.0q$, $-3.0q$, $+2.5q$, and $+4.0q$ are placed at the vertices of this rectangle, what is the electric potential at the rectangle's center? The value of q is 64 nC.

7. The largest premechanical earthwork was built about 700 years ago in Africa along the boundaries of the Benin Empire. The total length of the earthwork is estimated to be 16 000 km. Consider an equilateral triangle in space formed by three equally charged asteroids with charges of 7.2×10^{-2} C. If the length of each side of the triangle is 1.6×10^{4} km, what is the electric potential at the midpoint of any one of the sides?

8. The Royal Dragon restaurant in Bangkok, Thailand, can seat over 5000 customers, making it the largest restaurant in the world. The area of the restaurant is equal to that of a square with sides measuring 184 m each. If three charges, each equal to 25.0 nC, are placed at three vertices of this square, what is the electric potential at the fourth vertex?

NAME ______________________ DATE ____________ CLASS ____________

Electrical Energy and Current

Problem G

COST OF ELECTRICAL ENERGY

PROBLEM

In 1995, the fortune of Sir Muda Hassanal Bolkiah, Sultan of Brunei, was estimated at $37 billion. Suppose this money is used to pay for the energy used by an ordinary 1.0 kW microwave oven at a rate of $0.086/kW•h. How long can the microwave oven be powered?

SOLUTION

Given: Cost of energy = $0.086/kW•h $\quad P = 1.0$ kW

Purchase power = $\$37 \times 10^9$

Unknown: $\Delta t = ?$

First, calculate the number of kilowatt-hours that can be purchased by dividing the total amount of money ($) by the cost of energy ($/kW•h). Then divide the energy used (kW•h) by the power (kW) to find the time.

$$\text{Energy} = (\$37 \times 10^9)\left(\frac{1\ \text{kW}\bullet\text{h}}{\$0.086}\right) = 4.3 \times 10^{11}\ \text{kW}\bullet\text{h}$$

$$\Delta t = \frac{\text{Energy}}{P} = \frac{(4.3 \times 10^{11}\ \text{kW}\bullet\text{h})}{(1.0\ \text{kW})} = \boxed{4.3 \times 10^{11}\ \text{h} = 4.9\ \text{My}}$$

ADDITIONAL PRACTICE

1. The ten reactors of the nuclear power station in Fukushima, Japan, produce 8.8×10^9 W of electric power. If you have $\$1.0 \times 10^6$, how many hours of the station's energy output can you buy? Assume a price of $0.081/kW•h.
2. A power plant in Hawaii produces electricity by using the difference in temperature between surface water and deep-ocean water. The plant produces 104 kW. Suppose this energy is sold at a rate of $0.120/kW•h. For how long would $18 000 worth of this energy last?
3. For basketball fans, a flexible light source that can be attached around any basketball hoop rim has been developed. This source reportedly lasts for 1.0×10^4 h. How much power will this light source provide over its lifetime if its overall cost of operation is $23 and the energy cost is $0.086/kW•h?
4. Fluorescent lamps with resistances that can be adjusted from 80.0 Ω to 400.0 Ω are being produced. If such a lamp is connected to a 110 V emf source, how much will it cost to operate the lamp at its maximum rated power for 24 h? The cost of energy is $0.086/kW•h.
5. A solar-cell installation that can convert 15.5 percent of the sun's energy into electricity has been built. The device delivers 1.0 kW of power. If it produces energy at a cost of $0.080/kW•h, how much energy must the sun provide to produce $1000.00 worth of energy?

Circuits and Circuit Elements

Problem A

RESISTORS IN SERIES

PROBLEM

A particular electronic-code lock provides over 500 billion combinations. Moreover, it can sustain an electric shock of 1.25×10^5 V. Suppose this potential difference is applied across a series connection of the following resistors: 11.0 kΩ, 34.0 kΩ, and 215 kΩ. What is the equivalent resistance for the circuit? What current would pass through the resistors?

SOLUTION

1. DEFINE **Given:**

$$R_1 = 11.0\ \text{k}\Omega = 11.0 \times 10^3\ \Omega$$
$$R_2 = 34.0\ \text{k}\Omega = 34.0 \times 10^3\ \Omega$$
$$R_3 = 215\ \text{k}\Omega = 215 \times 10^3\ \Omega$$
$$\Delta V = 1.25 \times 10^5\ \text{V}$$

Unknown: $R_{eq} = ?$ $\quad I = ?$

2. PLAN **Choose the equation(s) or situation:** Because the resistors are in series, the equivalent resistance can be calculated from the equation for resistors in series.

$$R_{eq} = R_1 + R_2 + R_3$$

The equation relating potential difference, current, and resistance can be used to calculate the current.

$$\Delta V = IR_{eq}$$

Rearrange the equation(s) to isolate the unknown(s):

$$I = \frac{\Delta V}{R_{eq}}$$

3. CALCULATE **Substitute the values into the equation(s) and solve:**

$$R_{eq} = (11.0 \times 10^3\ \Omega + 34.0 \times 10^3\ \Omega + 215 \times 10^3\ \Omega)$$
$$R_{eq} = \boxed{2.60 \times 10^5\ \Omega}$$

$$I = \frac{(1.25 \times 10^5\ \text{V})}{(2.60 \times 10^5\ \Omega)} = \boxed{0.481\ \text{A}}$$

4. EVALUATE For resistors connected in series, the equivalent resistance should be greater than the largest resistance in the circuit.

$$2.60 \times 10^5\ \Omega > 2.15 \times 10^5\ \Omega$$

ADDITIONAL PRACTICE

1. The Large Optics Diamond Turning Machine at the Lawrence Livermore National Lab in California can cut a human hair lengthwise 3000 times! That would produce pieces with a cross-sectional area of 10^{-7} mm². A 1.0 m long silver wire with this cross-sectional area would have a resistance

of 160 kΩ. Consider three pieces of silver wire connected in series. If their lengths are 2.0 m, 3.0 m, and 7.5 m, and the resistance of each wire is proportional to its length, what is the equivalent resistance of the connection?

2. Most of the 43×10^3 kg of gold that sank with the British ship HMS *Laurentis* in 1917 has been recovered. If this gold were drawn into a wire long enough to wrap around Earth's equator five times, its electrical resistance would be about 5.0×10^8 Ω. Consider three resistors with resistances that are exactly 1/3, 2/7, and 1/5 the resistance of the gold wire. What equivalent resistance would be produced by connecting all three resistors in series?

3. A 3 mm thick steel wire that stretches for 5531 km has a resistance of about 82 kΩ. If you connect in series three resistors with the values 16 kΩ, 22 kΩ, and 32 kΩ, what value must the fourth resistor have for the equivalent resistance to equal 82 kΩ?

4. The largest operating wind generator in the world has a rotor diameter of almost 100 m. This generator can deliver 3.2 MW of power. Suppose you connect a 3.0 kΩ resistor, a 4.0 kΩ resistor, and a 5.0 kΩ resistor in series. What potential difference must be applied across these resistors in order to dissipate power equal to 1.00 percent of the power provided by the generator? What is the current through the resistors? (Hint: Recall the relation between potential difference, resistance, and power.)

5. The resistance of loudspeakers varies with the frequency of the sound they produce. For example, one type of speaker has a minimum resistance of 4.5 Ω at low frequencies and 4.0 Ω at ultra-low frequencies, and it has a peak resistance of 16 Ω at a high frequency. Consider a set of resistors with resistances equal to 4.5 Ω, 4.0 Ω, and 16.0 Ω. What values of the equivalent resistance can be obtained by connecting any two of them in different series connections?

6. Standard household potential difference in the United States is 1.20×10^2 V. However, many electric companies allow the residential potential difference to increase up to 138 V at night. Suppose a 2.20×10^2 Ω resistor is connected to a constant potential difference. A second resistor is provided in an alternate circuit so that when the potential difference rises to 138 V, the second resistor connects in series with the first resistor. This changes the resistance so that the current in the circuit is unchanged. What is the value of the second resistor?

7. The towers of the Golden Gate Bridge, in San Francisco, California, are about 227 m tall. The supporting cables of the bridge are about 90 cm thick. A steel cable with a length of 227 m and a thickness of 90 cm would have a resistance of 3.6×10^{-5} Ω. If a 3.6×10^{-5} Ω resistor is connected in series with an 8.4×10^{-6} Ω resistor, what power would be dissipated in the resistors by a 280 A current? (Hint: Recall the relation between current, resistance, and power.)

NAME ______________________ DATE ______________ CLASS ______________

Circuits and Circuit Elements

Problem B

RESISTORS IN PARALLEL

PROBLEM

A light bulb in a camper's flashlight is labeled 2.4 V, 0.70 A. Find the equivalent resistance and the current if three of these light bulbs are connected in parallel to a standard C size 1.5 V battery.

SOLUTION

1. DEFINE **Given:**

$\Delta V_1 = 2.4$ V $\quad I_1 = 0.70$ A

$\Delta V_2 = 2.4$ V $\quad I_2 = 0.70$ A

$\Delta V_3 = 2.4$ V $\quad I_3 = 0.70$ A

$\Delta V = 1.5$ V

Unknown: $R_{eq} = ?$ $\quad I = ?$

2. PLAN **Choose the equation(s) or situation:** Because the resistors (bulbs) are in parallel, the equivalent resistance can be calculated from the equation for resistors in parallel.

$$\frac{1}{R_{eq}} = \frac{1}{R_1} + \frac{1}{R_2} + \frac{1}{R_3}$$

To calculate the individual resistances, use the definition of resistance.

$$\Delta V_n = I_n R_n$$

The following form of the equation can be used to calculate the current.

$$\Delta V = I R_{eq}$$

Rearrange the equation(s) to isolate the unknown(s):

$$R_n = \frac{\Delta V_n}{I_n} \qquad I = \frac{\Delta V}{R_{eq}}$$

3. CALCULATE **Substitute the values into the equation(s) and solve:**

$$R_1 = \frac{\Delta V_1}{I_1} = \frac{(2.4\text{ V})}{(0.70\text{ A})} = 3.4\ \Omega$$

$$R_2 = \frac{\Delta V_2}{I_2} = \frac{(2.4\text{ V})}{(0.70\text{ A})} = 3.4\ \Omega$$

$$R_3 = \frac{\Delta V_3}{I_3} = \frac{(2.4\text{ V})}{(0.70\text{ A})} = 3.4\ \Omega$$

$$\frac{1}{R_{eq}} = \frac{1}{R_1} + \frac{1}{R_2} + \frac{1}{R_3} = \frac{3}{(3.4\ \Omega)} = \frac{0.88}{1\ \Omega}$$

$$R_{eq} = \boxed{1.1\ \Omega}$$

$$I = \frac{(1.5\text{ V})}{(1.1\ \Omega)} = \boxed{1.4\text{ V}}$$

4. EVALUATE For resistors connected in parallel, the equivalent resistance should be less than the smallest resistance in the circuit.

$$1.1\ \Omega < 3.4\ \Omega$$

ADDITIONAL PRACTICE

1. A certain full-range loudspeaker has a maximum resistance of 32 Ω at 45 Hz, a resistance of 5.0 Ω for most audible frequencies, and a resistance of only 1.8 Ω at 20 kHz. Consider three resistors with resistances of 1.8 Ω, 5.0 Ω, and 32 Ω. Find the equivalent resistance if they are connected in parallel.

2. The Large Electron Positron ring, near Geneva, Switzerland, is one of the biggest scientific instruments on Earth. The circumference of the ring is 27 km. A copper wire with this length and a cross-sectional area of 1 mm^2 will have a resistance of about 450 Ω. Consider a parallel connection of three resistors with resistances equal to 1.0, 2.0, and 0.50 times the resistance of the copper wire, respectively. What is the equivalent resistance?

3. Cars on the Katoomba Scenic Railway are pulled along by winding cables, and at one point, they move along a 310 m span that makes an angle of 51° with the horizontal. A 310 m steel cable that is 4 cm thick would have an estimated resistance of 2.48×10^{-2} Ω. An equivalent resistance of 6.00×10^{-3} Ω can be obtained if two resistors, one having the same resistance as the steel cable, are connected in parallel. Find the resistance of the second resistor.

4. In 1992 in Atlanta, 1 724 000 United States quarters were placed side by side in a straight line. Suppose these quarters were stacked to form a cylindrical tower. If the influence of the air gaps between coins is negligible, the resistance of the tower can be estimated easily. Find the resistance if the parallel connection of four resistors that have resistances equal to exactly 1, 3, 7, and 11 times the tower's resistance yields an equivalent resistance of 6.38×10^{-2} Ω.

5. The largest piece of gold ever found had a mass of about 70 kg. If you were to draw this mass of gold out into a thin wire with a cross-sectional area of 2.0 mm^2, the wire would have a length of 1813 km. The wire would also have a resistance per unit length of 1.22×10^{-2} Ω/m.
   a. What is the resistance of the wire?
   b. Suppose the wire were cut into pieces having resistance of exactly 1/2, 1/4, 1/5, and 1/20 of the wire's resistance, respectively. If these pieces are reconnected in parallel, what is the equivalent resistance of the four pieces?

6. A powerful cordless drill uses a 14.4 V battery to deliver 225 W of power. Treating the drill as a resistor, find its resistance. If a single 14.4 V battery is connected to four "drill" resistors that are connected in parallel, what are the equivalent resistance and the battery current?

7. The total length of the telephone wires in the Pentagon is 3.22×10^5 km. Suppose these wires have a resistance of 1.0×10^{-2} Ω/m. If all the wires are cut into 1.00×10^3 km pieces and all pieces are connected in parallel to a AA battery ($\Delta V = 1.50$ V), what would the current through the wires be? Assume that a AA battery can sustain this current.

Circuits and Circuit Elements

Problem C

EQUIVALENT RESISTANCE

PROBLEM

A certain amplifier can drive five channels with a load of 8.0 Ω each. Consider five 8.0 Ω resistors connected as shown. What is the equivalent resistance?

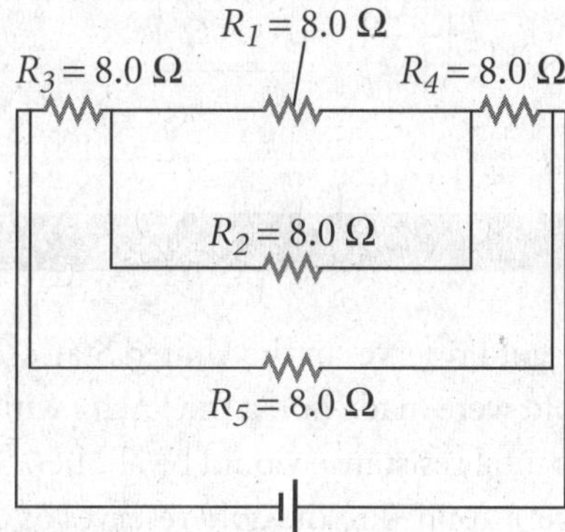

REASONING

Divide the circuit into groups of series and parallel resistors. This way, the methods presented in determining equivalent resistance for resistors in series and parallel can be used to calculate the equivalent resistance for each group.

SOLUTION

1. **Redraw the circuit as a group of resistors along one side of the circuit.** Bends in a wire do not affect the circuit and do not need to be represented in a schematic diagram. Redraw the circuit without corners, keeping the arrangement of the circuit elements the same and disregarding the emf source.

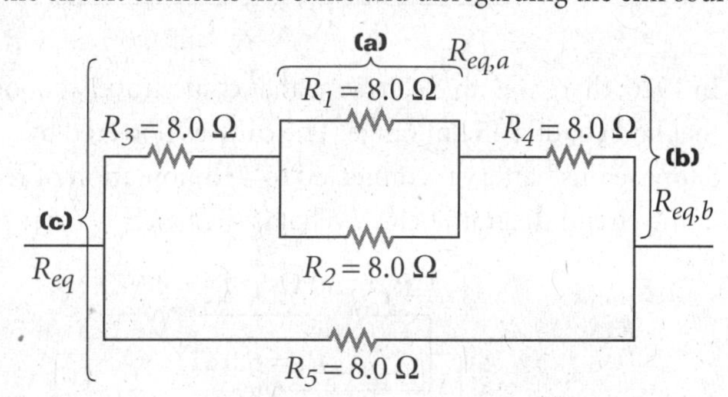

2. **Identify components in series, and calculate their equivalent resistance.** At this stage, there are no resistors in series.

3. **Identify components in parallel, and calculate their equivalent resistance.** Resistors in group (**a**) are in parallel. For group (**a**):

$$\frac{1}{R_{eq,a}} = \frac{1}{R_1} + \frac{1}{R_2} = \frac{2}{(8.0\ \Omega)} = \frac{0.25}{1\ \Omega}$$

$$R_{eq,a} = 4.0\ \Omega$$

4. Repeat steps 2 and 3 until the resistors in the circuit are reduced to a single equivalent resistance.

Resistors in group (**b**) are in series.

$$R_{eq,b} = R_{eq,a} + R_3 + R_4 = 4.0\ \Omega + 8.0\ \Omega + 8.0\ \Omega = 20.0\ \Omega$$

Resistors in group (**c**) are in parallel.

$$\frac{1}{R_{eq}} = \frac{1}{R_{eq,b}} + \frac{1}{R_5} = \frac{1}{(20.0\ \Omega)} + \frac{1}{(8.0\ \Omega)} = \frac{0.0500}{1\ \Omega} + \frac{0.12}{1\ \Omega}$$

$$R_{eq} = \left(\frac{0.17}{1\ \Omega}\right)^{-1} = \boxed{5.9\ \Omega}$$

ADDITIONAL PRACTICE

1. In 1993, the gold reserves in the United States were about 8.490×10^6 kg. If all that gold were made into a thick wire with a cross-sectional area of 1 cm^2, its total resistance would be about $6.60 \times 10^2\ \Omega$. If the same operation were applied to the gold reserves of Germany, France, and Switzerland, the resistances would be $2.40 \times 10^2\ \Omega$, $2.00 \times 10^2\ \Omega$, and $2.00 \times 10^2\ \Omega$, respectively. Now consider all four resistors connected as shown in the circuit diagram below. Find the equivalent resistance.

$R_1 = 6.60 \times 10^2\ \Omega$ $R_2 = 2.40 \times 10^2\ \Omega$

$R_4 = 2.00 \times 10^2\ \Omega$

$R_3 = 2.00 \times 10^2\ \Omega$

2. In 1920, there was an electric car that could travel at about 40 km/h and that had about a 45 km range. The car was powered by a 24 V battery. Suppose this battery is connected to a combination of resistors, as shown in the circuit diagram below. What is the battery current?

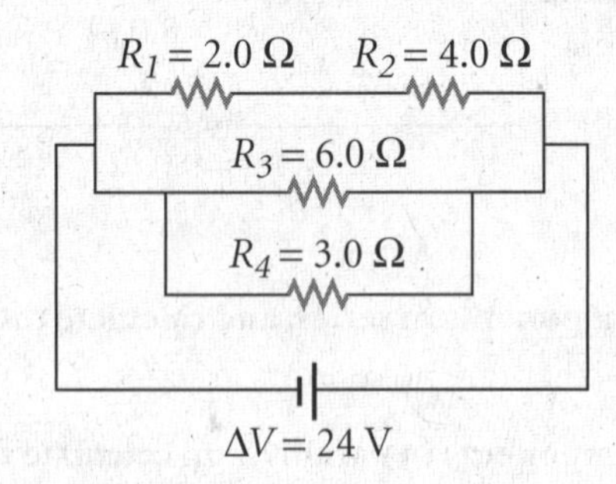

3. By adding water to the Enviro-Gen portable power pack, the device can generate 12 V for up to 40.0 h. If this device powers a combination of small appliances with the resistances shown in the circuit diagram below, what will be the net current for the circuit?

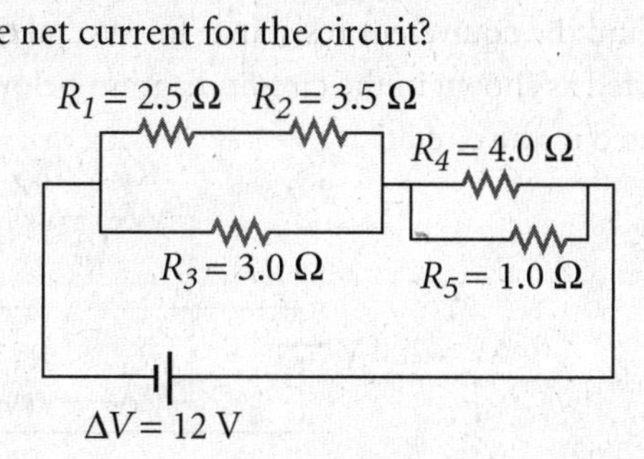

4. In 1995, the most powerful wind generator in the United States was the Z-40, which has a rotor diameter of 40 m. The machine is capable of producing 5.00×10^2 A at 1.00×10^3 V, assuming 100 percent efficiency. Suppose a direct current of 5.00×10^2 A is produced when a potential difference of 1.00×10^3 V is placed across a circuit of resistors, as shown in the diagram below. What is the equivalent resistance of the circuit? What is the power dissipated in the circuit?

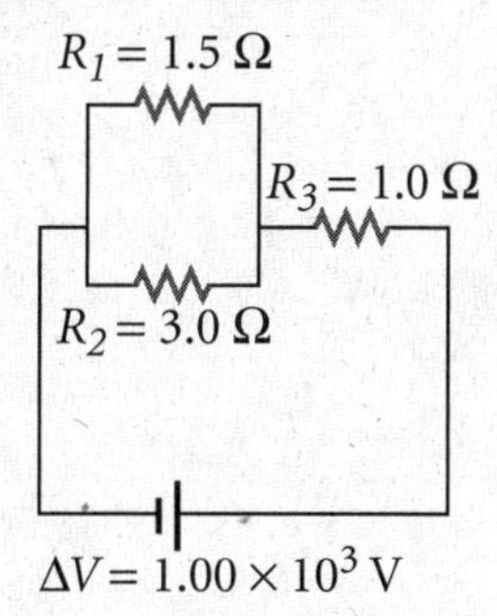

5. The longest-lasting battery in the world is at Oxford University, in England. It was built in 1840 and was still working in 1977, producing a 1.0×10^{-8} A current. The battery provided a potential difference of 2.00×10^3 V. If the battery is connected to a group of resistors, as shown in the circuit diagram below, find the value of the equivalent resistance and the value of *r*.

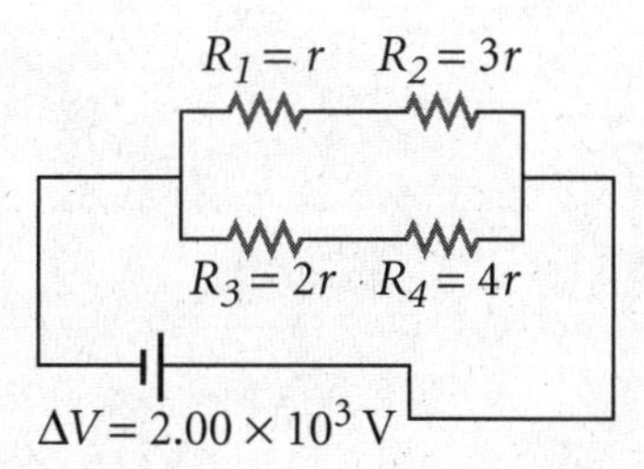

6. During World War II, a high-powered searchlight was produced that had a power rating of 6.0×10^5 W. Assuming a potential difference of 220 V across the searchlight, find the resistance of the light bulb in that searchlight. Find the equivalent resistance for several of these light bulbs connected as shown in the circuit diagram below. What is the total power dissipated in the circuit?

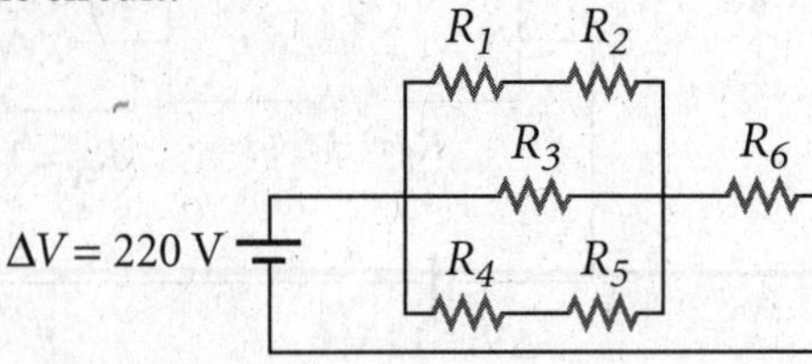

NAME ______________________ DATE ____________ CLASS ______________

Circuits and Circuit Elements

Problem D

CURRENT IN AND POTENTIAL DIFFERENCE ACROSS A RESISTOR

PROBLEM

For the circuit from the previous section's sample problem, determine the current in and potential difference across the 8.0 Ω resistor (R_4) in the figure below.

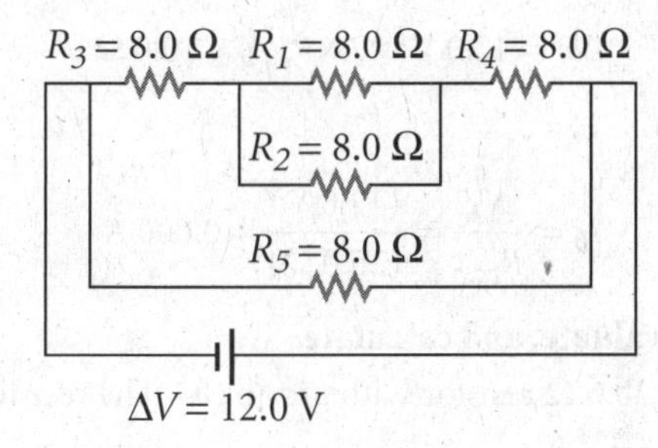

REASONING

First find the equivalent resistance of the circuit. From this, determine the total circuit current. Then rebuild the circuit in steps, calculating the current and potential difference for the equivalent resistance of each group until the current in and potential difference across the specified 8.0 Ω resistor are known.

SOLUTION

1. **Determine the equivalent resistance of the circuit.**
   The equivalent resistance, which was calculated in the sample problem of the previous section, is 5.7 Ω.

2. **Calculate the total current in the circuit.**
   Substitute the potential difference and equivalent resistance in $\Delta V = IR$, and rearrange the equation to find the current delivered by the battery.

$$I = \frac{\Delta V}{R_{eq}} = \frac{(12.0 \text{ V})}{(5.9\ \Omega)} = 2.0 \text{ A}$$

3. **Determine a path from the equivalent resistance found in step 1 to the specified resistor.**
   Review the path taken to find the equivalent resistance in the diagram below, and work backward through this path. The equivalent resistance for the entire circuit is the same as the equivalent resistance for group (**c**). The top resistors in group (**c**), in turn, form the equivalent resistance for group (**b**), and the rightmost resistor in group (**b**) is the specified 8.0 Ω resistor.

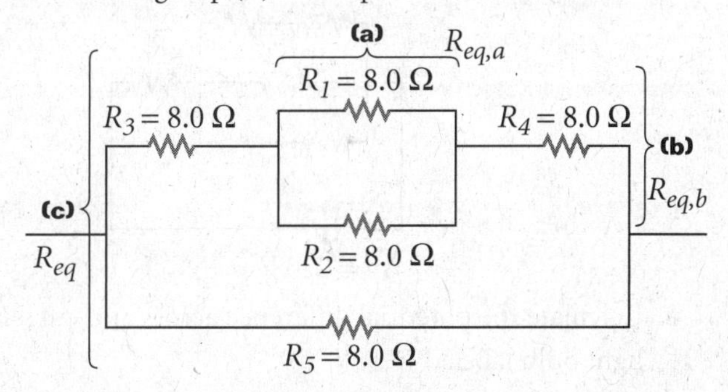

4. Follow the path determined in step 3, and calculate the current in and potential difference across each equivalent resistance. Repeat this process until the desired values are found.

Regroup, evaluate, and calculate.
Replace the circuit's equivalent resistance with group (**c**), as shown in the figure. The resistors in group (**c**) are in parallel, so the potential difference across each resistor is equal to the potential difference across the equivalent resistance, which is 12.0 V. The current in the equivalent resistance in group (**b**) can now be calculated using $\Delta V = IR$.

Given: $\Delta V = 12.0 \text{ V}$ $\quad R_{eq,b} = 20.0 \ \Omega$

Unknown: $I_b = ?$

$$I_b = \frac{\Delta V}{R_{eq,b}} = \frac{(12.0 \text{ V})}{(20.0 \ \Omega)} = 0.600 \text{ A}$$

Regroup, evaluate, and calculate.
Replace the 20.0 Ω resistor with group (**b**). The resistors R_3, $R_{eq,b}$, and R_4 in group (**b**) are in series, so the current in each resistor is the same as the current in the equivalent resistance, which equals 0.600 A.

$$I_b = \boxed{0.600 \text{ A}}$$

The potential difference across the 8.0 Ω resistor at the right can be calculated using $\Delta V = IR$.

Given: $I_b = 0.600 \text{ A}$ $\quad R_4 = 8.0 \ \Omega$

Unknown: $\Delta V = ?$

$$\Delta V = IR = (0.600 \text{ A})(8.0 \ \Omega) = \boxed{4.8 \text{ V}}$$

The current through the specified resistor is 0.600 A, and the potential difference across it is 4.8 V.

ADDITIONAL PRACTICE

1. Recall from the previous section the high-powered searchlight with the power rating of 6.0×10^5 W. For a potential difference of 220 V placed across the light bulb of this searchlight, you found a value for the bulb's resistance. You also determined the equivalent resistance for the circuit shown in the figure below.

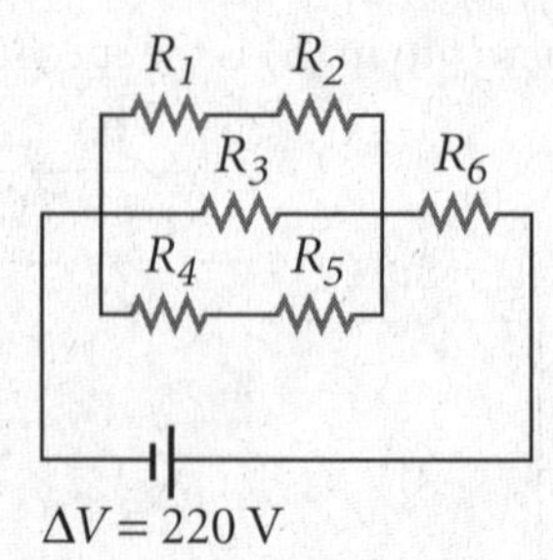

a. Calculate the potential difference across and current in the searchlight bulb labeled R_3.

b. Calculate the potential difference across and current in the search-light bulb labeled R_2.

c. Calculate the potential difference across and current in the search-light bulb labeled R_4.

2. Recall the portable power pack that can provide 12 V for 40.0 h. The device powers a combination of small appliances with the resistances shown in the circuit diagram below. In the previous section, you calculated the equivalent resistance and net current for this circuit.

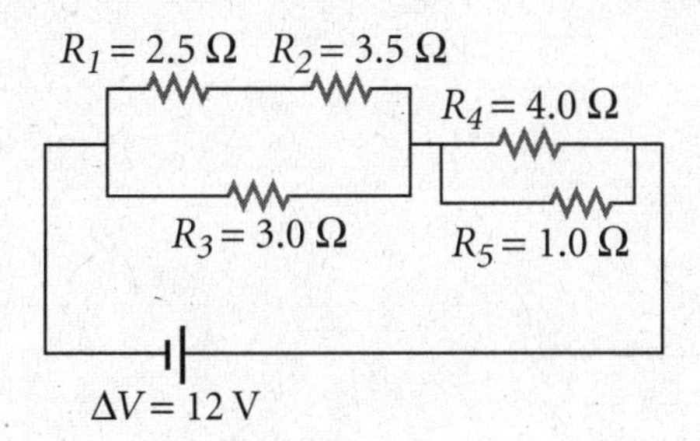

a. Calculate the potential difference across and current in the 1.0 Ω appliance.

b. Calculate the potential difference across and current in the 2.5 Ω appliance.

c. Calculate the potential difference across and current in the 4.0 Ω appliance.

d. Calculate the potential difference across and current in the 3.0 Ω appliance.

3. Recall the longest-lasting battery in the world, which was constructed at Oxford University in 1840. In 1977, the terminal voltage of the battery was 2.00×10^3 V. Suppose the battery is placed in the circuit shown in the diagram below. Determine the equivalent resistance of the circuit, and then find the following:

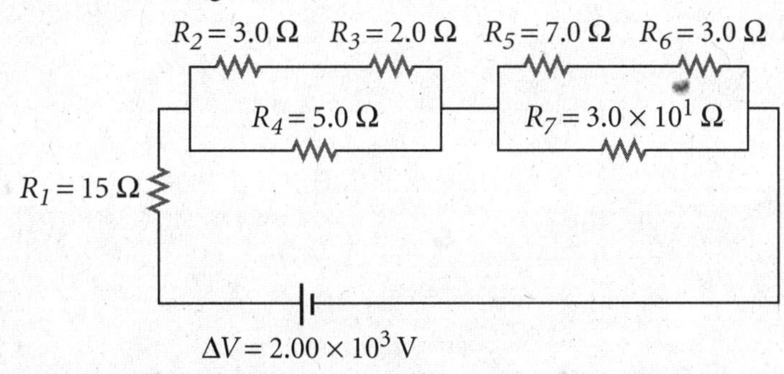

a. the potential difference and current in the 5.0 Ω resistor (R_4).

b. the potential difference and current in the 2.0 Ω resistor (R_3).

c. the potential difference and current in the 7.0 Ω resistor (R_5).

d. the potential difference and current in the 3.0×10^1 Ω resistor (R_7).

Magnetism

Problem A

PARTICLES IN A MAGNETIC FIELD

PROBLEM

A proposed shock absorber uses the properties of tiny magnetizable particles suspended in oil. In the presence of a magnetic field, the particles form chains, making the liquid nearly solid. Suppose this liquid is activated by a magnetic field with a magnitude of 0.080 T. What force will a particle with a charge of 2.0×10^{-11} C experience if it moves perpendicular to the magnetic field with a speed of 4.8 cm/s?

SOLUTION

Given: $q = 2.0 \times 10^{-11}$ C

$v = 4.8$ cm/s $= 4.8 \times 10^{-2}$ m/s

$B = 0.080$ T $= 8.0 \times 10^{-2}$ T $= 8.0 \times 10^{-2}$ N/A•m

Unknown: $F_{magnetic} = ?$

Use the equation for the magnitude of a magnetic field.

$$B = \frac{F_{magnetic}}{qv}$$

$$F_{magnetic} = qvB$$

$$F_{magnetic} = (2.0 \times 10^{-11}\ \text{C})(4.8 \times 10^{-2}\ \text{m/s})(8.0 \times 10^{-2}\ \text{N/A}\bullet\text{m})$$

$$F_{magnetic} = \boxed{7.7 \times 10^{-14}\ \text{N}}$$

ADDITIONAL PRACTICE

1. In 1995, construction began on the world's most powerful electromagnet, which will be capable of producing a magnetic field with a strength of 45 T. If an electron enters this field at a right angle with a speed of 7.5×10^6 m/s, what will be the magnetic force acting on the electron?
2. Transrapid, the world's first train using magnetic field levitation, is designed to glide 1.0 cm above the track at speeds of up to 450 km/h. Suppose a charged particle with a charge of 12×10^{-9} C and a speed of 450 km/h moves at right angles to a 2.4 T magnetic field. What is the magnetic force acting on the particle?
3. Europe's fastest train can move at speeds of up to 350 km/h. What is the magnetic force on a particle with a 36 nC charge that travels 350 km/h at a 30.0° angle with respect to Earth's magnetic field? Assume the strength of Earth's magnetic field is 7.0×10^{-5} T. (Hint: Recall that a magnetic field affects only the velocity component perpendicular to the field.)
4. The U.S. Navy plans to use electromagnetic catapults to launch planes from aircraft carriers. These catapults will be able to accelerate a 45×10^3 kg plane to a launch speed of 260 km/h. If an electron travels at a

speed of 2.60×10^2 km/h perpendicular to a magnetic field, so that the force acting on the electron is 3.0×10^{-17} N, what is the magnetic field strength?

5. A solar-powered vehicle developed at Massachusetts Institute of Technology can reach an average speed of 60.0 km/h. Due to the presence of Earth's magnetic field, a magnetic force acts on the electrons that move through the car's circuitry. Suppose an electron moves at 60.0 km/h perpendicular to Earth's magnetic field. If a 2.0×10^{-22} N force acts on the electron, what is the magnetic field strength at that location?

6. By 1905, a locomotive powered by an electric motor (which was in turn powered by an internal-combustion engine) cruised between the East Coast and West Coast. Suppose a particle with a charge of 88×10^{-9} C moves at the same speed as the locomotive in a 0.32 T magnetic field. If the magnetic force on the particle is 1.25×10^{-6} N, what is the particle's (and locomotive's) speed?

7. NASA plans a shuttle launch system, called MagLifter, that would speed up the spacecraft using superconducting magnets. A space shuttle would be accelerated along a track 4.0 km long. At the end of the track, the speed of the shuttle would be great enough to place it into orbit around Earth.

a. At what speed is an electron moving in a powerful 6.4 T magnetic field if it experiences a force of 2.76×10^{-16} N?

b. If this is the shuttle's final speed at the end of the track, how long does it take the shuttle to glide along the track, assuming that the shuttle undergoes constant acceleration from rest?

8. The mass spectrometer is a device that was first used to separate the different isotopes of an element. In a mass spectrometer, ionized atoms with the same speed enter a region in which a strong magnetic field is maintained. The field, which is perpendicular to the plane in which the atoms move, exerts a force that keeps the atoms in a circular path. Because all of the atoms have the same charge and speed, the magnetic force exerted on them is the same. However, the atoms have slightly different masses, and therefore the centripetal acceleration on each varies slightly. The centripetal force depends on the speed of the atoms and the radius of the circular path the atoms follow. Isotopes with larger masses have trajectories with larger radii. As a result, the more-massive isotopes are detected farther from the center of the apparatus than isotopes with smaller masses.

a. Suppose the magnetic field in a mass spectrometer has a field strength of 0.600 T and that two lithium atoms, each with a single positive charge of 1.60×10^{-19} C and a speed of 2.00×10^5 m/s, enter that field. What is the magnetic force exerted on these two atoms?

b. If the masses of the atoms are 9.98×10^{-27} kg and 11.6×10^{-27} kg, respectively, what will be the difference in the radius for each atom's path? (Hint: The magnetic force equals the force required to keep the atoms in a circular path.)

Magnetism

Problem B

FORCE ON A CURRENT-CARRYING CONDUCTOR

PROBLEM

In March 1967, a 1000 T magnetic field was obtained in a lab for a fraction of a second. Maximum sustained fields presently do not exceed 50 T. If a 25 cm long wire is perpendicular to a 1.0×10^3 T magnetic field and the magnetic force on the wire is 1.6×10^2 N, what is the current in the wire?

SOLUTION

Given: $B = 1.0 \times 10^3$ T

$F_{magnetic} = 1.6 \times 10^2$ N

$\ell = 25$ cm $= 0.25$ m

Unknown: $I = ?$

Use the equation for force on a current-carrying conductor perpendicular to a magnetic field.

$$I = \frac{F_{magnetic}}{B\ell}$$

$$I = \frac{1.6 \times 10^2 \text{ N}}{(1.0 \times 10^3 \text{ T})(0.25 \text{ m})} = \boxed{0.64 \text{ A}}$$

ADDITIONAL PRACTICE

1. In 1964, a magnet at the Francis Bitter National Magnet Laboratory created a magnetic field with a magnitude of 22.5 T. Ten megawatts of power was required to generate this field. If a wire that is 12 cm long and that carries a current of 8.4×10^{-2} A is placed in this field at a right angle to the field, what is the force acting on it?
2. One of the subway platforms in downtown Chicago is 1066 m long. Suppose the current in one of the service wires below the platform is 0.80 A. What is the magnitude of Earth's magnetic field at that location if the field exerts a force of 6.3×10^{-2} N on the wire? Assume the current is perpendicular to the field.
3. The distance between the pylons of the power line across the Ameralik Fjord in Greenland is 5.376 km. If this length of wire carries a current of 12 A and experiences a force of 3.1 N because of Earth's magnetic field, what is the magnitude of the field in that region? Assume the wire makes an angle of 38° with the field. (Hint: Only the component of the magnetic field that is perpendicular to the current contributes to the magnetic force.)
4. In February 1996, NASA extended a 21.0 km conducting tether from the space shuttle *Columbia* in order to see how much power could be generated by interacting with Earth's magnetic field. Suppose the

magnetic field at the shuttle's location has a magnitude of 6.40×10^{-7} T. What current must be induced in the tether, located perpendicular to the field, to create a magnetic force of 1.80×10^{-2} N?

5. Magnetic fields are created by the currents in home appliances. Suppose the magnetic field in the vicinity of an electric blow-dryer has a magnitude of about 2.5×10^{-4} T. If a wire 4.5 cm long in such a blow-dryer experiences a magnetic force of 3.6×10^{-7} N, what current must exist in the wire? Assume the wire is perpendicular to the field.

6. At Sandia National Laboratories, in Albuquerque, New Mexico, an interesting project has been developed. Engineers have proposed a train that will roll on *existing* rails and use *existing* wheels but be propelled by magnetic forces. Suppose such a train is pushed by a 5.0×10^{5} N force and results from the interaction between a current in many wires and a magnetic field with a magnitude of 3.8 T that is oriented perpendicular to the wires. Find the total length of the wires if they carry a 2.00×10^{2} A current.

7. The largest electric sign used for advertising was set on the Eiffel Tower, in Paris. It was lighted by 250 000 bulbs, which required an immense number of electric cables. Find the total length of those cables if a single straight cable of identical length experiences a force of 16.1 N in Earth's magnetic field. Assume the magnitude of the magnetic field is 6.4×10^{-5} T and the wire, carrying a current of 2.8 A, is placed perpendicular to the field.

8. Cows have four stomachs, so if a cow swallows a nail, it is hard to remove it. To extract the nail, a veterinarian gives the cow a strong magnet to swallow. Suppose that such a "cow magnet" has a magnetic field magnitude of 0.040 T. If a 55 cm straight length of wire located near the magnet carries a current of 0.10 A and is positioned so that the angle between the magnetic field lines and the current is 45°, what is the magnetic force that the current will experience? (Hint: Only the component of the magnetic field that is perpendicular to the current contributes to the magnetic force.)

9. For many years, the strongest magnetic field ever produced had a magnitude of 38 T. Suppose a straight wire with a length of 2.0 m is located perpendicular to this field. What current would have to pass through the wire in order for the magnetic force to equal the weight of a graduate student with a mass of 75 kg?

10. The longest straight span of railroad tracks stretches 478 km in southwestern Australia. Suppose an electric current is sent through one of the rails and that a force of 0.40 N is exerted by Earth's magnetic field. If the magnetic field has a strength of 7.50×10^{-5} T at that location and is perpendicular to the rail, how large is the current?

NAME ______________________ DATE ______________ CLASS ______________

Electromagnetic Induction

Problem A

INDUCED emf AND CURRENT

PROBLEM

In 1994, a unicycle with a wheel diameter of 2.5 cm was ridden 3.6 m in Las Vegas, Nevada. Suppose the wheel has 200 turns of thin wire wrapped around its rim, creating loops with the same diameter as the wheel. An emf of 9.6 mV is induced when the wheel is perpendicular to a magnetic field that steadily decreases from 0.68 T to 0.24 T. For how long is the emf induced?

SOLUTION

1. DEFINE

Given:

$N = 200$ turns

$D = 2.5 \text{ cm} = 2.5 \times 10^{-2} \text{ m}$

$B_i = 0.68 \text{ T}$

$B_f = 0.24 \text{ T}$

$\text{emf} = 9.6 \text{ mV} = 9.6 \times 10^{-3} \text{ V}$

$\theta = 0.0°$

Unknown: $\Delta t = ?$

2. PLAN

Choose the equation(s) or situation: Use Faraday's law of magnetic induction to find the induced emf in the coil. Only the magnetic field strength changes with time.

$$\text{emf} = -N\frac{\Delta\theta \text{m}}{\Delta t} = -\frac{N\Delta[AB\cos\theta]}{\Delta t} = -NA\cos\theta\frac{\Delta B}{\Delta t}$$

Use the equation for the area of a circle to calculate the area (A).

$$A = \pi r^2 = \pi\left(\frac{D}{2}\right)^2$$

Rearrange the equation(s) to isolate the unknown(s):

$$\Delta t = -NA\cos\theta\frac{\Delta B}{\text{emf}}$$

3. CALCULATE

Substitute values into the equation(s) and solve:

$$A = \pi\left(\frac{2.5 \times 10^{-2} \text{ m}}{2}\right)^2 = 4.9 \times 10^{-4} \text{ m}^2$$

$$\Delta t = -(200)(4.9 \times 10^{-4} \text{ m}^2)[\cos 0.0°]\frac{(0.24 \text{ T} - 0.68 \text{ T})}{(9.6 \times 10^{-3} \text{ V})}$$

$$\Delta t = \boxed{4.5 \text{ s}}$$

4. EVALUATE

The induced emf is directed through the coiled wire so that the magnetic field produced opposes the decrease in the applied magnetic field. The rate of this change is indicated by the positive value of time.

NAME ______________________ DATE ____________ CLASS ____________

ADDITIONAL PRACTICE

1. The Pentagon covers an area of 6.04×10^5 m^2, making it one of the world's largest office buildings. Suppose a huge loop of wire is placed on the ground so that it covers the same area as the Pentagon. If the loop is pulled at opposite ends so that its area decreases, an emf will be induced because of the component of Earth's magnetic field that is perpendicular to the ground. If the field component has a strength of 6.0×10^{-5} T and the average induced emf in the loop is 0.80 V, how much time passes before the loop's area is reduced by half?
2. A Japanese-built Ferris wheel has a diameter of 100.0 m and can carry almost 500 people at a time. Suppose a huge magnet is used to create a field with an average strength of 0.800 T perpendicular to the wheel. The magnet is then pulled away so that the field steadily decreases to zero over time. If the wheel is a single conducting circular loop and the induced emf is 46.7 V, find the time needed for the magnetic field to decrease to zero.
3. In 1979, a potential difference of about 32.0 MV was measured in a lab in Tennessee. The maximum magnetic fields, obtained for very brief periods of time, had magnitudes around 1.00×10^3 T. Suppose a coil with exactly 50 turns of wire and a cross-sectional area of 4.00×10^{-2} m^2 is placed perpendicular to one of these extremely large magnetic fields, which quickly drops to zero. If the induced emf is 32.0 MV, in how much time does the magnetic field strength decrease from 1.00×10^3 T to 0.0 T?
4. The world's largest retractable roof is on the SkyDome in Toronto, Canada. Its area is 3.2×10^4 m^2, and it takes 20 min for the roof to fully retract. If you have a coil with exactly 300 turns of wire that changes its area from 0.0 m^2 to 3.2×10^4 m^2 in 20.0 min, what will be the emf induced in the coil? Assume that a uniform perpendicular magnetic field with a strength of 4.0×10^{-2} T passes through the coil.
5. At Massachusetts Institute of Technology, there is a specially shielded room in which extremely weak magnetic fields can be measured. As of 1994, the smallest field measured had a magnitude of 8.0×10^{-15} T. Suppose a loop having an unknown number of turns and an area equal to 1.00 m^2 is placed perpendicular to this field and the magnitude of the field strength is increased tenfold in 3.0×10^{-2} s. If the emf induced is equal to -1.92×10^{-11} V, how many turns are in the loop?
6. An electromagnet that has a mass of almost 8.0×10^6 kg was built at the CERN particle physics research facility in Switzerland. As part of the detector in one of the world's largest particle accelerators, this magnet creates a fairly large magnetic field with a magnitude of 0.50 T. Consider a coil of wire that has 880 equal turns. Suppose this loop is placed perpendicular to the magnetic field, which is gradually decreased to zero in 12 s. If an emf of 147 V is induced, what is the area of the coil?

Electromagnetic Induction

Problem B

rms CURRENT AND emf

PROBLEM

In 1945, turbo-electric trains in the United States were capable of speeds exceeding 160 km/h. Steam turbines powered the electric generators, which in turn powered the driving wheels. Each generator produced enough power to supply an rms potential difference of 6.0×10^3 V across an 18 Ω resistor. What was the maximum potential difference across the resistor? What was the maximum current in the resistor? What was the rms current in the resistor? What was the generator's power output?

SOLUTION

1. DEFINE **Given:** $\Delta V_{rms} = 6.0 \times 10^3$ V $\quad R = 18\ \Omega$

Unknown: $\Delta V_{max} = ?$ $\quad I_{max} = ?$ $\quad I_{rms} = ?$ $\quad P = ?$

2. PLAN **Choose the equation(s) or situation:** Use the equation relating maximum and rms potential differences to calculate the maximum potential difference. Use the definition of resistance to calculate the maximum current, then use the equation relating maximum and rms currents to calculate rms current. Power can be calculated from the product of rms current and rms potential difference.

$$\Delta V_{max} = \sqrt{2}(\Delta V_{rms})$$

$$I_{max} = \frac{\Delta V_{max}}{R}$$

$$I_{rms} = \frac{I_{max}}{\sqrt{2}}$$

$$P = \Delta V_{rms} I_{rms}$$

3. CALCULATE **Substitute values into the equation(s) and solve:**

$$\Delta V_{max} = (1.41)(6.0 \times 10^3 \text{ V})$$

$$= \boxed{8.5 \times 10^3 \text{ V}}$$

$$I_{max} = \frac{(8.5 \times 10^3 \text{ V})}{(18\ \Omega)}$$

$$= \boxed{4.7 \times 10^2 \text{ A}}$$

$$I_{rms} = (0.707)(4.7 \times 10^2 \text{ A})$$

$$= \boxed{3.3 \times 10^2 \text{ A}}$$

$$P = (6.0 \times 10^3 \text{ V})(3.3 \times 10^2 \text{ A})$$

$$= \boxed{2.0 \times 10^6 \text{ W}}$$

4. EVALUATE To determine whether severe rounding errors occurred through the various calculations, obtain the product of the maximum current and maximum potential difference. The product of ΔV_{max} and I_{max} should equal $2P$, which for this problem equals 4.0×10^6 W.

ADDITIONAL PRACTICE

1. In 1963, the longest single-span “rope way” for cable cars opened in California. The rope way stretched about 4 km from the Coachella Valley to Mount San Jacinto. Suppose the rope way, which is actually a steel cable, becomes icy. To de-ice the cable, you can connect its two ends to a 120 V (rms) generator. If the resistance of the cable is 6.0×10^{-2} Ω,
   a. what will the rms current in the cable be?
   b. what will the maximum current in the cable be?
   c. what power will be dissipated by the cable, thus melting the ice?
2. In 1970, a powerful sound system was set up on the Ontario Motor Speedway in California to make announcements to more than 200 000 people over the noise of 50 racing cars. The *acoustic* power of that system was 30.8 kW. If the system was driven by a generator that provided an rms potential difference of 120.0 V and only 10.0 percent of the supplied power was transformed into acoustic power, what was the maximum current in the sound system?
3. Modern power plants typically have outputs of over 10×10^6 kW. But in 1905, the Ontario Power Station, built on the Niagara River, produced only 1.325×10^5 kW. Consider a single generator producing this power when it is connected to a single load (resistor). If the generated rms potential difference is 5.4×10^4 V, what is the maximum current and the value of the resistor?
4. Stability of potential difference is a major concern for all high-emf sources. In 1996, James Cross of the University of Waterloo in Canada, constructed a compact power supply that produces a *stable* potential difference of 1.024×10^6 V. It can provide 2.9×10^{-2} A at this potential difference. If these values are the rms quantities for an alternating current source, what are the maximum potential difference and current?
5. Certain species of catfish found in Africa have “power plants” similar to those of electric eels. Though the electricity generated is not as powerful as that of some eels, the electric catfish can discharge 0.80 A with a potential difference of 320 V. Consider an ac generator in a circuit with a load. If its maximum values for potential difference and current are the same as the potential difference and current for the catfish, what are the rms values for the potential difference and current? What is the resistance of the load?
6. A wind generator installed on the island of Oahu, Hawaii, has a rotor that is about 100 m in diameter. When the wind is strong enough, the generator can produce a maximum current of 75 A in a 480 Ω load. Find the rms potential difference across the load.
7. The world’s first commercial tidal power plant, built in France, has a power output of only 6.2×10^4 kW, produced by 24 generators. Find the power produced by each generator. If one of these generators is connected to a 120 kΩ resistor, find the rms and maximum currents in it.

Electromagnetic Induction

Problem C

TRANSFORMERS

PROBLEM

The span between the pylons for the power line serving the Buksefjorden Hydro Power Station, in Greenland, is about 5 km. These power lines carry electricity at a high potential difference, which is then stepped-down to the standard European household potential difference of 220 V. If the transformer that does this has a primary with 1.5×10^5 turns and a secondary with 250 turns, what is the potential difference across the primary? (Note: Reduction of potential difference usually takes place over several steps, not one.)

SOLUTION

1. DEFINE

Given: $N_1 = 1.5 \times 10^5$ turns $\quad N_2 = 250$ turns

$\Delta V_2 = 220$ V

Unknown: $\Delta V_1 = ?$

2. PLAN

Choose the equation(s) or situation: Use the transformer equation.

$$\frac{\Delta V_2}{\Delta V_1} = \frac{N_2}{N_1}$$

Rearrange the equation(s) to isolate the unknown(s):

$$\Delta V_1 = \Delta V_2 \frac{N_1}{N_2}$$

3. CALCULATE

Substitute values into the equation(s) and solve:

$$\Delta V_1 = (220 \text{ V})\frac{(1.5 \times 10^5 \text{ turns})}{(250 \text{ turns})}$$

$$\Delta V_1 = \boxed{1.3 \times 10^5 \text{ V}}$$

4. EVALUATE

The primary should have 130 kV. The step-down factor for the transformer is 600:1.

ADDITIONAL PRACTICE

1. The most powerful electromagnet in the world uses 24 MW of power. Consider a transformer that transmits 24 MW of power. The primary has 5600 turns, and the secondary has 240 turns. If the secondary potential difference is 4.1 kV, what is the primary potential difference?

2. In 1990, New Jersey was the most densely populated state in the United States, with 403 people per square kilometer. Consider a transformer with 74 turns in the primary and 403 turns in the secondary. If a 650 V potential difference exists between the terminals of the secondary, what is the potential difference between the terminals of the primary?

3. In 1992, a battery whose longest dimension was 70 nm was made at the University of California at Irvine. It produced a 2.0×10^{-2} V potential difference for almost an hour. Suppose an ac generator producing this potential difference is connected to a transformer that contains exactly 400 turns in the primary and exactly 3600 turns in the secondary. If this emf is created between the terminals of the primary of a step-up transformer, what is the potential difference between the terminals of the secondary? If the 2.0×10^{-2} V is applied to the secondary instead, what is the potential difference created between the terminals of the primary?

4. The hydraulic turbines installed at the third power plant at Grand Coulee, in Washington, are almost 10 m in diameter. Each turbine is large enough to produce a current of 1.0×10^{3} A at 765 kV. If a transformer steps this potential difference down to 540 kV and the primary contains 2.8×10^{3} turns, how many turns must be in the secondary?

5. In 1965, the biggest power failure to date left about 30 million people in the dark for several hours. About 200 000 km^2, including Ontario, Canada, and several northeastern states in the United States were affected. Such failures can usually be avoided because all major electric grids are interconnected. Transformers are needed in these connections. For example, a transformer can increase the potential difference from 230 kV to 345 kV, which is the typical potential difference for transmission lines in the United States. If the primary in such a step-up transformer has 1.2×10^{4} turns, how many turns are in the secondary?

6. The sunroof on some cars doubles as a solar battery. In strong sunlight, it produces about 20.0 W of power.

a. If this power is transmitted by the primary of a transformer with 120 V across it, what is the current in the primary?

b. What is the potential difference across the secondary if the primary contains only 36 percent of the number of turns in the secondary?

7. Electric cars, though still few in number, are appearing on the roads. By using a 220 V potential difference to recharge their batteries instead of the standard 120 V, the cars could be recharged in 3 to 6 h. One process, developed at the Electric Power Research Institute in California, suggests using 220 V outlets with a 30.0 A charging current. If a transformer is used to increase the potential difference from 120 V to 220 V, what is the current in the primary? How many turns does the primary have if the secondary has 660 turns?

Atomic Physics

Problem A

QUANTUM ENERGY

PROBLEM

Free-electron lasers can be used to produce a beam of light with variable wavelength. Because the laser can produce light with wavelengths as long as infrared waves or as short as X rays, its potential applications are much greater than for a laser that can produce light of only one wavelength. If such a laser produces photons of energies ranging from 1.034 eV to 620.6 eV, what are the minimum and the maximum wavelengths corresponding to these photons?

SOLUTION

Given:

$$E_1 = 1.034 \text{ eV}$$
$$= (1.034 \text{ eV})\left(1.60 \times 10^{-19} \frac{\text{J}}{\text{eV}}\right) = 1.65 \times 10^{-19} \text{ J}$$
$$E_2 = 620.6 \text{ eV}$$
$$= (620.6 \text{ eV})\left(1.60 \times 10^{-19} \frac{\text{J}}{\text{eV}}\right) = 9.93 \times 10^{-17} \text{ J}$$
$$h = 6.63 \times 10^{-34} \text{ J}\bullet\text{s}$$
$$c = 3.00 \times 10^8 \text{ m/s}$$

Unknown: $\lambda_{min} = ?$ $\lambda_{max} = ?$

Use the equation for the energy of a quantum of light. Use the relationship between the frequency and wavelength of electromagnetic waves.

$$E = hf$$
$$f = \frac{c}{\lambda}$$

Substitute for f in the first equation, and rearrange to solve for wavelength.

$$E = \frac{hc}{\lambda}$$
$$\lambda = \frac{hc}{E}$$

Substitute values into the equation.

$$\lambda_{max} = \frac{(6.63 \times 10^{-34} \text{ J}\bullet\text{s})(3.00 \times 10^8 \text{ m/s})}{(1.65 \times 10^{-19} \text{ J})}$$
$$\lambda_{max} = 1.21 \times 10^{-6} \text{ m}$$
$$\lambda_{max} = \boxed{1210 \text{ nm}}$$

$$\lambda_{min} = \frac{(6.63 \times 10^{-34} \text{ J}\bullet\text{s})(3.00 \times 10^8 \text{ m/s})}{(9.93 \times 10^{-17} \text{ J})}$$
$$\lambda_{min} = 2.00 \times 10^{-9} \text{ m}$$
$$\lambda_{min} = \boxed{2.00 \text{ nm}}$$

NAME ______________________ DATE ______________ CLASS ______________

ADDITIONAL PRACTICE

1. In 1974, IBM researchers announced that X rays with energies of 1.29×10^{-15} J had been guided through a "light pipe" similar to optic fibers used for visible and near-infrared light. Calculate the wavelength of one of these X-ray photons.

2. Some strains of *Mycoplasma* are the smallest living organisms. The wavelength of a photon with 6.6×10^{-19} J of energy is equal to the length of one *Mycoplasma*. What is that wavelength?

3. Of the various types of light emitted by objects in space, the radio signals emitted by cold hydrogen atoms in regions of space that are located between stars are among the most common and important. These signals occur when the "spin" angular momentum of an electron in a hydrogen atom changes orientation with respect to the "spin" angular momentum of the atom's proton. The energy of this transition is equal to a fraction of an electron-volt, and the photon emitted has a very low frequency. Given that the energy of these radio signals is 5.92×10^{-6} eV, calculate the wavelength of the photons.

4. The camera with the fastest shutter speed in the world was built for research with high-power lasers and can expose individual frames of film with extremely high frequency. If the frequency is the same as that of a photon with 2.18×10^{-23} J of energy, calculate its magnitude.

5. Wireless "cable" television transmits images using radio-band photons with energies of around 1.85×10^{-23} J. Find the frequency of these photons.

6. In physics, the basic units of measurement are based on fundamental physical phenomena. For example, one second is defined by a certain transition in a cesium atom that has a frequency of *exactly* 9 192 631 770 s^{-1}. Find the energy in electron-volts of a photon that has this frequency. Use the unrounded values for Planck's constant ($h = 6.626\,0755 \times 10^{-34}$ J•s) and for the conversion factor between joules and electron volts (1 eV = $1.602\,117\,33 \times 10^{-19}$ J).

7. Consider an electromagnetic wave that has a wavelength equal to 92 cm, a length that corresponds to the longest ear of corn grown to date. What is the frequency corresponding to this wavelength? What is its photon energy? Express the answer in joules and in electron-volts.

8. The slowest machine in the world, built for testing stress corrosion, can be controlled to operate at speeds as low as 1.80×10^{-17} m/s. Find the distance traveled at this speed in 1.00 year. Calculate the energy of the photon with a wavelength equal to this distance.

Atomic Physics

Problem B

THE PHOTOELECTRIC EFFECT

PROBLEM

Ultraviolet radiation, which is part of the solar spectrum, causes a photoelectric effect in certain materials. If the kinetic energy of the photoelectrons from an aluminum sample is 5.6×10^{-19} J and the work function of aluminum is about 4.1 eV, what is the frequency of the photons that produce the photoelectrons?

SOLUTION

Given: $KE_{max} = 5.6 \times 10^{-19}\ \text{J} = 3.5\ \text{eV}$

$hf_t = 4.1\ \text{eV}$

$h = 6.63 \times 10^{-34}\ \text{J}\bullet\text{s} = 4.14 \times 10^{-15}\ \text{eV}\bullet\text{s}$

$c = 3.00 \times 10^{8}\ \text{m/s}$

Unknown: $f = ?$

Use the equation for the maximum kinetic energy of an ejected photoelectron to calculate the frequency of the photon.

$$KE_{max} = \frac{hc}{\lambda} - hf_t$$

$$f = \frac{KE_{max} + hf_t}{h}$$

$$f = \frac{[3.5\ \text{eV} + 4.1\ \text{eV}]}{4.14 \times 10^{-15}\ \text{eV/s}}$$

$$f = \boxed{1.8 \times 10^{15}\ \text{Hz}}$$

ADDITIONAL PRACTICE

1. The melting point of mercury is about −39°C, which makes it convenient for many applications, such as thermometers and thermostats. Mercury also has some unusual applications, such as in “liquid mirrors.” By spinning a pool of mercury, a perfect parabolic surface can be obtained for use as a concave mirror. If the surface of mercury is exposed to light, a photoelectric effect can be observed. If the work function of mercury is 4.5 eV, what is the frequency of photons that produce photoelectrons with kinetic energies of 3.8 eV?

2. The largest all-metal telescope mirror, which was used in Lord Rosse’s telescope, the Leviathan, was produced in 1845 from a copper-tin alloy. The work function of the surface of that mirror can be estimated as 4.3 eV. Calculate the frequency of the photons that would produce photoelectrons having a kinetic energy of 3.2 eV.

3. Values for the work function have been experimentally determined for most nonradioactive, elemental metals. The smallest work function, which is 2.14 eV, belongs to the element cesium. The largest work function, which is 5.9 eV, belongs to the element selenium.
   a. What is the wavelength of the photon that will just have the threshold energy for cesium?
   b. What is the wavelength of the photon that will just have the threshold energy for selenium?

4. Carbon is a nonmetal, yet it is a conductor of electricity, and it exhibits photoelectric properties. Calculate the work function and the threshold frequency for carbon if photons with a wavelength of 2.00×10^2 nm produce photoelectrons moving at a speed of 6.50×10^5 m/s.

5. Two giant water jugs made in 1902 for the Maharaja of Jaipur, India, are the largest single-piece silver items on Earth. Each jug has a capacity of about 8 m^3 and a mass of more than 240 kg. If the surface of one of these jugs is exposed to UV light that has a frequency of 2.2×10^{15} Hz, a photoelectric effect is observed. If the photoelectrons have 4.4 eV of kinetic energy, find the work function and the threshold frequency of silver.

6. Rhenium is one of the rarest elements on Earth. Estimates indicate that on average there is less than 1 μg of rhenium in a kilogram of Earth's crust and about 4 ng of rhenium in a liter of sea water. Rhenium also has one of the highest work functions of any metal. Suppose that a rhenium sample is exposed to light with a wavelength of 2.00×10^2 nm and that photoelectrons with kinetic energies of 0.46 eV are emitted. Using this information, calculate the work function and threshold frequency for rhenium.

7. Sodium-vapor lamps, which are widely used in streetlights, produce yellow light with a principal wavelength of 589 nm. Would this sodium light cause a photoelectric effect on the surface of solid sodium ($hf_t = 2.3$ eV)? If the answer is yes, what is the energy of those photoelectrons? If the answer is no, how much energy does the photon lack?

8. Lithium's unusual electric properties make it an ideal material for high-capacity batteries. The purity of a thin lithium foil, used in a lithium-polymer "sandwich" to create an efficient battery for solar-powered cars, is extremely important. One way to assess a metal's purity is by means of the photoelectric effect. If the work function of lithium is 2.3 eV, what is the kinetic energy of the photoelectrons produced by violet light with a wavelength of 410 nm?

9. Lead and zinc are vital elements in the construction of electric batteries. For example, the largest battery in the world, in use in California, is a lead-acid battery, while the most durable battery in the world, working continuously since 1840, is a zinc-sulfur battery. Zinc and lead have similar work functions: 4.3 eV and 4.1 eV, respectively. Suppose certain photons have just enough energy to cause a photoelectric effect in zinc. If the same photons were to strike the surface of lead, what would be the speed of the photoelectrons?

NAME ______________________ DATE ____________ CLASS ______________

Atomic Physics

Problem C

INTERPRETING ENERGY-LEVEL DIAGRAMS

PROBLEM

Although neutral lithium has three electrons, two of the electrons are in the $n = 1$ energy level, which is filled. The third electron is in the $n = 2$ energy level, and behaves like the single electron in a hydrogen atom, except that the nucleus has a +3 charge and the two inner electrons partially shield the stronger attraction of the nucleus. Of the various possible energy levels that can be occupied by one or more of lithium's electrons, the simplest one for the single outer electron is shown below.

E_4 ———————— $E = 4.53$ eV

E_3 ———————— $E = 3.84$ eV

E_2 ———————— $E = 1.85$ eV

E_1 ———————— $E = 0$ eV

An electron in a lithium atom drops from an energy level to E_1. If the photon emitted has a wavelength of 323.7 nm, which energy level did the electron initially occupy, and what is its value (in eV) relative to E_1? Assume that the ground state E_1 has an energy of 0 eV.

SOLUTION

1. DEFINE **Given:** $\lambda = 323.7$ nm

$E_{final} = E_1 = 0$ eV

Unknown: $E_{initial} = ?$

2. PLAN **Choose the equation(s) or situation:** Calculate the energy of the emitted photon, and equate this to the difference between the energy levels.

$$E = hf$$

$$c = f\lambda$$

$$E = \frac{hc}{\lambda}$$

$$E = E_{initial} - E_{final} = E_{initial} - E_1$$

$$E_{initial} = E + E_1 = \frac{hc}{\lambda} + E_1$$

3. CALCULATE **Substitute the values into the equation(s) and solve:**

$$E_{initial} = \frac{(6.63 \times 10^{-34}\ \text{J•s})(3.00 \times 10^{8}\ \text{m/s})}{323.7\ \text{nm}} \times \frac{10^{9}\ \text{nm}}{1\ \text{m}} \times \frac{1\ \text{eV}}{1.60 \times 10^{-19}\ \text{J}} + 0\ \text{eV}$$

$$E_{initial} = 3.84\ \text{eV} + 0\ \text{eV} = \boxed{3.84\ \text{eV}}$$

4. EVALUATE The photon drops from the energy level that is 3.84 eV higher than the ground state E_1, which according to the energy diagram is the level with energy E_3. It is this transition in the lithium atom that causes the emission of the photon with wavelength 323.7 nm.

ADDITIONAL PRACTICE

1. Using the energy diagram for lithium shown in the Sample problem, determine which energy level provides the initial state for an electron that, in dropping to energy E_1, produces photon with the wavelength 671.9 nm.

2. Aluminum has electrons that fill the orbitals of the two lowest energy states, and another pair of electrons that fill the 3*s* level. This leaves one unpaired electron in the 3*p* that can behave in certain transitions like the single electron of the hydrogen atom. A diagram of some of these transitions is shown below.

E_4 ———————— $E = 5.24$ eV

E_3 ———————— $E = 4.69$ eV

E_2 ———————— $E = 3.15$ eV

E_1 ———————— $E = 0$ eV

If an electron drops from the E_4 energy level to E_1, what is the wavelength of the emitted photon?

3. Calculate the wavelength of a photon emitted between the E_3 and E_1 energy levels of an aluminum atom.

4. Calculate the wavelength of a photon emitted between the E_2 and E_1 energy levels of an aluminum atom.

NAME ______________________ DATE ______________ CLASS ______________

Atomic Physics

Problem D

DE BROGLIE WAVES

PROBLEM

In 1974, the most massive elementary particle known was the ψ' particle, which has a mass of about four times the mass of a proton. If the de Broglie wavelength is 3.615×10^{-11} m when it has a speed of 2.80×10^3 m/s, what is the particle's mass?

SOLUTION

Given: $\lambda = 3.615 \times 10^{-11}$ m $\quad v = 2.80 \times 10^3$ m/s

$h = 6.63 \times 10^{-34}$ J•s

Unknown: $m = ?$

Use the equation for the de Broglie wavelength.

$$m = \frac{h}{\lambda v} = \frac{(6.63 \times 10^{-34}\ \text{J•s})}{(3.615 \times 10^{-11}\ \text{m})(2.80 \times 10^3\ \text{m/s})} = \boxed{6.55 \times 10^{-27}\ \text{Kg}}$$

ADDITIONAL PRACTICE

1. The world's smallest watch, made in Switzerland, has a fifteen-jewel mechanism and is less than 5 mm wide. When this watch has a speed of 3.2 m/s, its de Broglie wavelength is 3.0×10^{-32} m. What is the mass of the watch?
2. Discovered in 1983, Z° was still the most massive particle known in 1995. If the de Broglie wavelength of the Z° particle is 6.4×10^{-11} m when the particle has a speed of 64 m/s, what is the particle's mass?
3. Although beryllium, Be, is toxic, the Be^{2+} ion is harmless. When a Be^{2+} ion is accelerated through a potential difference of 240 V, the ion's de Broglie wavelength is 4.4×10^{-13} m. What is the mass of the Be^{2+} ion?
4. In 1972, a powder with a particle size of 2.5 nm was produced. At what speed should a neutron move to have a de Broglie wavelength of 2.5 nm?
5. The graviton is a hypothetical particle that is believed to be responsible for gravitational interactions. Although its existence has not been proven, cosmological observations and theories indicate that its mass, which is theoretically zero, has an upper limit of 7.65×10^{-70} kg. What speed must a graviton have for its de Broglie wavelength to be 5.0×10^{32} m? (Gravitons are predicted to have a speed equal to that of light.)
6. The average mass of the bee hummingbird is about 1.6 g. What is the de Broglie wavelength of this variety of hummingbird if it is flying at 3.8 m/s?
7. In 1990, Dale Lyons ran 42 195 m, in 3 h, 47 min, while carrying a spoon with an egg in it. What was Lyons' average speed during the run? If the egg's mass was 0.080 kg, what was its de Broglie wavelength?

NAME ______________________ DATE ______________ CLASS ______________

Subatomic Physics

Problem A

BINDING ENERGY

PROBLEM

Each egg of *Caraphractus cinctus*, a parasitic wasp, has a mass of about 2.0×10^{-10} kg. Consider the formation of $^{16}_{8}O$ from H atoms and neutrons. How many nuclei of $^{16}_{8}O$ must be formed to produce a mass defect equal to the mass of one *Caraphractus cinctus* egg? What is the total binding energy of these $^{16}_{8}O$ nuclei? The atomic mass of $^{16}_{8}O$ is 15.994 915 u.

SOLUTION

1. DEFINE

Given:

$m_{egg} = 2.0 \times 10^{-10}$ kg

element formed $= {}^{16}_{8}O$

$Z = 8$ $\qquad N = 16 - 8 = 8$

atomic mass of $^{16}_{8}O = 15.994\ 915$ u

atomic mass of H $= 1.007\ 825$ u

$m_n = 1.008\ 665$ u

Unknown:

$\Delta m = ?$

n = number of $^{16}_{8}O$ nuclei formed = ?

total $E_{bind} = ?$

2. PLAN

Choose the equation(s) or situation: First find the mass defect using the equation for mass defect.

$$\Delta m = Z(\text{atomic mass of H}) + Nm_n - \text{atomic mass}$$

To determine the number of oxygen-16 nuclei that must be formed to produce a total mass defect equal to the mass of a *Caraphractus cinctus* egg, calculate the ratio of the mass of one egg to the mass defect.

$$\text{the number of } {}^{16}_{8}O \text{ nuclei formed} = n = \frac{m_{egg}}{\Delta m}$$

Finally, to find the total binding energy of all $^{16}_{8}O$ nuclei, use the equation for the binding energy of a nucleus and multiply it by n.

$$\text{total } E_{bind} = nE_{bind} = n\Delta mc^2$$

$$\text{total } E_{bind} = n\Delta m\,(931.49 \text{ MeV/u})$$

3. CALCULATE

Substitute the values into the equation(s) and solve:

$$\Delta m = 8(1.007\ 825 \text{ u}) + 8(1.008\ 665 \text{ u}) - 15.994\ 915 \text{ u}$$

$$\Delta m = 8.062\ 600 \text{ u} + 8.069\ 320 \text{ u} - 15.994\ 915 \text{ u}$$

$$\Delta m = 0.137\ 005 \text{ u}$$

$$n = \frac{(2.0 \times 10^{-10} \text{ kg})}{(0.137\ 005 \text{ u})}\left(\frac{1 \text{ u}}{1.66 \times 10^{-27} \text{ kg}}\right) = \boxed{8.8 \times 10^{17} \text{ nuclei}}$$

$$\text{total } E_{bind} = (8.8 \times 10^{17})(0.137\ 005 \text{ u})(931.49 \text{ MeV/u})$$

$$\text{total } E_{bind} = \boxed{1.1 \times 10^{20} \text{ MeV}}$$

4. EVALUATE For every nucleus of $^{16}_{8}O$ formed, the mass defect is 0.137 005 u, and the mass of the nucleus formed is 15.994 915 u. When Δm has a total value of 2.0×10^{-10} kg, 8.8×10^{17} nuclei, or 2.3×10^{-8} kg of $^{16}_{8}O$ will have formed.

ADDITIONAL PRACTICE

1. In 1993, the United States had more than 100 operational nuclear reactors producing about 30 percent of the world's nuclear energy, or 610 TW•h.

a. Find the mass defect corresponding to a binding energy equal to that energy output.

b. How many $^{2}_{1}H$ nuclei would be needed to provide this mass defect?

c. How many $^{56}_{26}Fe$ nuclei would be needed to provide this mass defect?

d. How many $^{226}_{88}Ra$ nuclei would be needed to provide this mass defect?

2. In 1976, Montreal hosted the Summer Olympics. To complete the new velodrome, the 4.1×10^{7} kg roof had to be raised 10.0 cm to be placed in the exact position.

a. Find the energy needed to raise the roof.

b. Find the mass of $^{56}_{26}Fe$ that is formed when an amount of energy equal to that calculated in part (a) is obtained from binding H atoms and neutrons in iron-56 nuclei.

3. Nuclear-energy production worldwide was 2.0×10^{3} TW•h in 1993. What mass of $^{235}_{92}U$ releases an equivalent amount of energy in the form of binding energy?

4. In 1993, the United States burned about 2.00×10^{8} kg of coal to produce about 2.1×10^{19} J of energy. Suppose that instead of *burning* coal, you obtain energy by *forming* coal ($^{12}_{6}C$) out of H atoms and neutrons. What amount of coal must be formed to provide 2.1×10^{19} J of energy? Assume 100 percent efficiency.

5. The sun radiates energy at a rate of 3.9×10^{26} J/s. Suppose that all the sun's energy occurs because of the formation of $^{4}_{2}He$ from H atoms and neutrons. Find the number of reactions that take place each second.

6. Sulzer Brothers, a Swiss company, made powerful diesel engines for the container ships built for American President Lines. The power of each 12-cylinder engine is about 42 MW. Suppose the turbines use the formation of $^{14}_{7}N$ for the energy-releasing process. What mass of nitrogen would have to be formed to provide enough energy for 24 h of continuous work? Assume the turbines are 100 percent efficient.

7. A hundred years ago, the most powerful hydroelectric plant in the world produced 3.84×10^{7} W of electric power. Find the total mass of $^{12}_{6}C$ atoms that must be formed each second from H atoms and neutrons to produce the same power output.

NAME ______________________ DATE ____________ CLASS ____________

Subatomic Physics

Problem B

NUCLEAR DECAY

PROBLEM

The most stable radioactive nuclide known is tellurium-128. It was discovered in 1924, and its radioactivity was proven in 1968. This isotope undergoes two-step beta decay. Write the equations that correspond to this reaction.

SOLUTION

Given: $^{128}_{52}\text{Te} \rightarrow {}^{0}_{-1}e + \text{X} + \overline{\nu}$

$\text{X} \rightarrow {}^{0}_{-1}e + \text{Z} + \overline{\nu}$

Unknown: the daughter elements X and Z

The mass numbers and atomic numbers on both sides of the expression must be the same so that both charge and mass are conserved during the course of this particular decay reaction.

$$\text{Mass number of X} = 128 - 0 = 128$$

$$\text{Atomic number of X} = 52 - (-1) = 53$$

The periodic table shows that the nucleus with an atomic number of 53 is iodine, I. Thus, the first step of the process is as follows:

$$^{128}_{52}\text{Te} \rightarrow {}^{0}_{-1}e + {}^{128}_{53}\text{I} + \overline{\nu}$$

A similar approach for the second beta decay reaction gives the following equation. Again, the emission of an electron does not change the mass number of the nucleus. It does, however, change the atomic number by 1.

$$\text{Mass number of Z} = 128 - 0 = 128$$

$$\text{Atomic number of Z} = 53 - (-1) = 54$$

The periodic table shows that the nucleus with an atomic number of 54 is xenon, Xe. Thus the next step of the process is as follows:

$$^{128}_{53}\text{I} \rightarrow {}^{0}_{-1}e + {}^{128}_{54}\text{Xe} + \overline{\nu}$$

The complete two-step reaction is described by the two balanced equations below.

$$^{128}_{52}\text{Te} \rightarrow {}^{0}_{-1}e + {}^{128}_{53}\text{I} + \overline{\nu}$$

$$^{128}_{53}\text{I} \rightarrow {}^{0}_{-1}e + {}^{128}_{54}\text{Xe} + \overline{\nu}$$

ADDITIONAL PRACTICE

1. Standard nuclear fission reactors use $^{235}_{92}\text{U}$ for fuel. However, the supply of this uranium isotope is limited. Its concentration in natural uranium-238 is low, and the cost of enrichment is high. A good alternative is the breeder reactor in which the following reaction sequence occurs: $^{238}_{92}\text{U}$ captures a neutron, and the resulting isotope emits a $^{0}_{-1}e$ particle to form

$^{239}_{93}Np$. This nuclide emits a second $^{0}_{-1}e$ particle to form $^{239}_{94}Pu$, which is fissionable and can be used as an energy-producing material. Write balanced equations for each of the reactions described.

2. Radon has the highest density of any gas. Under normal conditions radon's density is about 10 kg/m^3. One of radon's isotopes undergoes two alpha decays and then one beta decay (β^-) to form $^{212}_{83}Bi$. Write the equations that correspond to these reaction steps.

3. Every element in the periodic table has isotopes, and cesium has the most: as of 1995, 37 isotopes of cesium had been identified. One of cesium's most stable isotopes undergoes beta decay (β^-) to form $^{135}_{56}Ba$. Write the equation describing this beta-decay reaction.

4. Fission is the process by which a heavy nucleus decomposes into two lighter nuclei and releases energy. Uranium-235 undergoes fission when it captures a neutron. Several neutrons are produced in addition to the two light daughter nuclei. Complete the following equations, which describe two types of uranium-235 fission reactions.

$$^{235}_{92}U + ^{1}_{0}n \rightarrow ^{144}_{56}Ba + ^{89}_{36}Kr + \_\_\_\_$$

$$^{235}_{92}U + ^{1}_{0}n \rightarrow ^{140}_{54}Xe + \_\_\_\_ + 2^{1}_{0}n$$

5. The maximum safe amount of radioactive thorium-228 in the air is 2.4×10^{-19} kg/m^3, which is equivalent to about half a kilogram distributed over the entire atmosphere. One reason for this substance's high toxicity is that it undergoes alpha decay in which gamma rays are produced as well. Write the equation corresponding to this reaction.

6. The 1930s were notable years for nuclear physics. In 1931, Robert Van de Graaff built an electrostatic generator that was capable of creating the high potential differences needed to accelerate charged particles. In 1932, Ernest O. Lawrence and M. Stanley Livingston built the first cyclotron. In the same year, Ernest Cockcroft and John Walton observed one of the first artificial nuclear reactions. Complete the following equation for the nuclear reaction observed by Cockcroft and Walton.

$$^{1}_{1}p + ^{7}_{3}Li \rightarrow \_\_\_\_ + ^{4}_{2}He$$

7. Among the naturally occurring elements, astatine is the least abundant, with less than 0.2 g present in Earth's entire crust. The isotope $^{217}_{85}At$ accounts for only about 5×10^{-9} g of all astatine. However, this highly radioactive isotope contributes nothing to the natural abundance of astatine because when it is created, it immediately undergoes alpha decay. Write the equation for this decay reaction.

NAME ______________________ DATE ____________ CLASS ____________

Subatomic Physics

Problem C

MEASURING NUCLEAR DECAY

PROBLEM

Tellurium-128, the most stable of all radioactive nuclides, has a half-life of about 1.5×10^{24} years. Determine the decay constant of $^{128}_{52}Te$. How long would it take for 75 percent of a sample of this isotope to decay?

SOLUTION

1. DEFINE **Given:** $T_{1/2} = 1.5 \times 10^{24}$ year

$N = 1.00 - 0.75 = 0.25 = 25$ percent remaining

Unknown: $\lambda = ?$ $t = ?$

2. PLAN **Choose the equation (s) or situation:** To calculate the decay constant, use the equation for determining a half-life.

$$T_{1/2} = \frac{0.693}{\lambda}$$

After 75 percent of the original sample decays, 25 percent, or one-fourth, of the parent nuclei remain. The rest of the nuclei have decayed into daughter nuclei. By definition, after the elapsed time of one half-life, half of the parent nuclei remain. After two half-lives, one fourth (or half of one-half) of the parent nuclei remain. The time for 75 percent of the original sample to decay is therefore two half-lives.

$$t = 2(T_{1/2})$$

Rearrange the equation(s) to isolate the unknown(s):

$$\lambda = \frac{0.693}{T_{1/2}}$$

3. CALCULATE **Substitute the values into the equation(s) and solve:**

$$\lambda = \frac{0.693}{(1.5 \times 10^{24}\ \text{year})} = \boxed{4.6 \times 10^{-24}\ \text{year}^{-1}}$$

$$t = 2(1.5 \times 10^{24}\ \text{year}) = \boxed{3.0 \times 10^{24}\ \text{year}}$$

4. EVALUATE Because the half-life of tellurium-128 is on the order of 10^{24} years, the decay constant is on the order of the reciprocal of the half-life, or 10^{-24} year. This corresponds to one decay event per year for a mole (127.6 g) of pure tellurium-128.

ADDITIONAL PRACTICE

1. In 1995, an Ethiopian athlete, Halie Gebrselassie, ran 10 km in 26 min, 43.53 s. Lead-214 has a half-life similar to Gebrselassie's run time. Find the decay constant for $^{214}_{82}Pb$. What percent of a sample of this isotope would decay in a time interval equal to five times Gebrselassie's run time?

2. Thorium-228, the most toxic of radioactive substances, has a half-life of 1.91 years. How long would it take for a sample of this isotope to decrease its toxicity by 93.75 percent?

3. In 1874, a huge cloud of locusts covered Nebraska, occupying an area of almost 5.00×10^5 km^2. The number of insects in that cloud was estimated as 10^{13}. Consider a sample of $^{144}_{56}Ba$, which has a half-life of 11.9 s, containing 1.00×10^{13} atoms. Approximately how long would it take for 1.25×10^{12} atoms to remain?

4. The oldest living tree in the world is a bristlecone pine in California named Methuselah. Its estimated age is 4800 years. Suppose a sample of $^{226}_{88}Ra$ began to decay at the time the pine began to grow. What percent of the sample would remain now? The half-life for $^{226}_{88}Ra$ is 1600 years.

5. The oldest fish on record is an eel that lived to the age of 88 years in a museum aquarium in Sweden. After that period of time, only about $\frac{1}{16}$ of a sample of $^{214}_{82}Pb$ would be left. Find the decay constant for $^{210}_{82}Pb$.

6. In 1994, Peter Hogg sailed across the Pacific Ocean on his trimaran *Aotea* in 34 days, 6 h, and 26 min. During that time period, only about $\frac{1}{512}$ of a sample of radon-222 would not decay. Estimate the decay constant of $^{222}_{86}Rn$ in (days)$^{-1}$.

7. Lithium-5 is the least stable isotope known. Its mean lifetime is 4.4×10^{-22} s. Use the definition of a radionuclide's mean lifetime as the reciprocal of the decay constant ($T = 1/\lambda$) to calculate the half-life of lithium-5.

NAME ______________________ DATE ____________ CLASS ____________

Advanced Topics

Problem A

ANGULAR DISPLACEMENT

PROBLEM

The diameter of the outermost planet, Pluto, is just 2.30×10^3 km. If a space probe were to orbit Pluto near the planet's surface, what would the arc length of the probe's displacement be after it had completed eight orbits?

SOLUTION

Given: $r = \frac{2.30 \times 10^3 \text{ km}}{2} = 1.15 \times 10^3 \text{ km}$

$\Delta\theta = 8 \text{ orbits} = 8(2\pi \text{ rad}) = 16\pi \text{ rad}$

Unknown: $\Delta s = ?$

Use the angular displacement equation and rearrange to solve for Δs.

$$\Delta\theta = \frac{\Delta s}{r}$$

$$\Delta s = r\Delta\theta = (1.15 \times 10^3 \text{ km})(16\pi \text{ rad}) = \boxed{5.78 \times 10^4 \text{ km}}$$

ADDITIONAL PRACTICE

1. A neutron star can have a mass three times that of the sun and a radius as small as 10.0 km. If a particle travels +15.0 rad along the equator of a neutron star, through what arc length does the particle travel? Does the particle travel in the clockwise or counterclockwise direction from the viewpoint of the neutron star's "north" pole?

2. John Glenn was the first American to orbit Earth. In 1962, he circled Earth counterclockwise three times in less that 5 h aboard his spaceship *Friendship 7*. If his distance from the Earth's center was 6560 km, what arc length did Glenn and his spaceship travel through?

3. Jupiter's diameter is 1.40×10^5 km. Suppose a space vehicle travels along Jupiter's equator with an angular displacement of 1.72 rad.

a. Through what arc length does the space vehicle move?

b. How many orbits around Earth's equator can be completed if the vehicle travels at the surface of the Earth through this arc length around Earth? Use 6.37×10^3 km for Earth's radius.

4. In 1981, the space shuttle *Columbia* became the first reusable spacecraft to orbit Earth. The shuttle's total angular displacement as it orbited Earth was 225 rad. How far from Earth's center was *Columbia* if it moved through an arc length of 1.50×10^6 km? Use 6.37×10^3 km for Earth's radius.

5. Mercury, the planet closest to the sun, has an orbital radius of 5.8×10^7 km. Find the angular displacement of Mercury as it travels through an arc length equal to the radius of Earth's orbit around the sun (1.5×10^8 km).

6. From 1987 through 1988, Sarah Covington-Fulcher ran around the United States, covering a distance of 1.79×10^4 km. If she had run that distance clockwise around Earth's equator, what would her angular displacement have been? (Earth's average radius is 6.37×10^3 km).

NAME ______________________ DATE ______________ CLASS ______________

Advanced Topics

Problem B

ANGULAR VELOCITY

PROBLEM

In 1975, an ultrafast centrifuge attained an average angular speed of 2.65 × 10^4 rad/s. What was the centrifuge's angular displacement after 1.5 s?

SOLUTION

Given: $\omega_{avg} = 2.65 \times 10^4$ rad/s

$\Delta t = 1.5$ s

Unknown: $\Delta\theta = ?$

Use the angular speed equation and rearrange to solve for $\Delta\theta$.

$$\omega_{avg} = \frac{\Delta\theta}{\Delta t}$$

$$\Delta\theta = \omega_{avg}\Delta t = (2.65 \times 10^4 \text{ rad/s})(1.5 \text{ s})$$

$$\Delta\theta = \boxed{4.0 \times 10^4 \text{ rad}}$$

ADDITIONAL PRACTICE

1. The largest steam engine ever built was constructed in 1849. The engine had one huge cylinder with a radius of 1.82 m. If a beetle were to run around the edge of the cylinder with an average angular speed of 1.00×10^{-1} rad/s, what would its angular displacement be after 60.0 s? What arc length would it have moved through?

2. The world's largest planetarium dome is 30 m in diameter. What would your angular displacement be if you ran around the perimeter of this dome for 120 s with an average angular speed of 0.40 rad/s?

3. The world's most massive magnet is located in a research center in Dubna, Russia. This magnet has a mass of over 3.0×10^7 kg and a radius of 30.0 m. If you were to run around this magnet so that you traveled 5.0×10^2 m in 120 s, what would your average angular speed be?

4. A floral clock in Japan has a radius of 15.5 m. If you ride a bike around the clock, making 16 revolutions in 4.5 min, what is your average angular speed?

5. A revolving globe with a radius of 5.0 m was built between 1982 and 1987 in Italy. If the globe revolves with the same average angular speed as Earth, how long will it take for a point on the globe's equator to move through 0.262 rad?

6. A water-supply tunnel in New York City is 1.70×10^2 km long and has a radius of 2.00 m. If a beetle moves around the tunnel's perimeter with an average angular speed of 5.90 rad/s, how long will it take the beetle to travel a distance equal to the tunnel's length?

Advanced Topics

Problem C

ANGULAR ACCELERATION

PROBLEM

A self-propelled Catherine wheel (a spinning wheel with fireworks along its rim) with a diameter of 20.0 m was built in 1994. Its maximum angular speed was 0.52 rad/s. How long did the wheel undergo an angular acceleration of 2.6 × 10^{-2} rad/s^2 in order to reach its maximum angular speed?

SOLUTION

Given:

$\omega_1 = 0$ rad/s

$\omega_2 = 0.52$ rad/s

$\alpha_{avg} = 2.6 \times 10^{-2}$ rad/s^2

Unknown: $\Delta t = ?$

Use the angular acceleration equation and rearrange to solve for Δt.

$$\alpha_{avg} = \frac{\Delta\omega}{\Delta t}$$

$$\Delta t = \frac{\Delta\omega}{\alpha_{avg}} = \frac{\omega_2 - \omega_1}{\alpha_{avg}} = \frac{0.52 \text{ rad/s} - 0 \text{ rad/s}}{2.6 \times 10^{-2} \text{ rad/s}^2} = \boxed{2.0 \times 10^1 \text{ s}}$$

ADDITIONAL PRACTICE

1. Peter Rosendahl of Sweden rode a unicycle with a wheel diameter of 2.5 cm. If the wheel's average angular acceleration was 2.0 rad/s^2, how long would it take for the wheel's angular speed to increase from 0 rad/s to 9.4 rad/s?
2. Jupiter has the shortest day of all of the solar system's planets. One rotation of Jupiter occurs in 9.83 h. If an average angular acceleration of -3.0×10^{-8} rad/s^2 slows Jupiter's rotation, how long does it take for Jupiter to stop rotating?
3. In 1989, Dave Moore of California built the Frankencycle, a bicycle with a wheel diameter of more than 3 m. If you ride this bike so that the wheels' angular speed increases from 2.00 rad/s to 3.15 rad/s in 3.6 s, what is the average angular acceleration of the wheels?
4. In 1990, David Robilliard rode a bicycle on the back wheel for more than 5 h. If the wheel's initial angular speed was 8.0 rad/s and Robilliard tripled this speed in 25 s, what was the average angular acceleration?
5. Earth takes about 365 days to orbit once around the sun. Mercury, the innermost planet, takes less than a fourth of this time to complete one revolution. Suppose some mysterious force causes Earth to experience an average angular acceleration of 6.05×10^{-13} rad/s^2, so that after

12.0 days its angular orbital speed is the same as that of Mercury. Calculate this angular speed and the period of one orbit.

6. The smallest ridable tandem bicycle was built in France and has a length of less than 40 cm. Suppose this bicycle accelerates from rest so that the average angular acceleration of the wheels is 0.800 rad/s^2. What is the angular speed of the wheels after 8.40 s?

Advanced Topics

Problem D

ANGULAR KINEMATICS

PROBLEM

In 1990, a pizza with a radius of 18.7 m was baked in South Africa. Suppose this pizza was placed on a rotating platform. If the pizza accelerated from rest at 5.00 rad/s^2 for 25.0 s, what was the pizza's final angular speed?

SOLUTION

Given: $\omega_i = 0$ rad/s

$\alpha = 5.00$ rad/s^2

$\Delta t = 25.0$ s

Unknown: $\omega_f = ?$

Use the rotational kinematic equation relating final angular speed to initial angular speed, angular acceleration, and time.

$$\omega_f = \omega_i + \alpha \Delta t$$

$$\omega_f = 0 \text{ rad/s} + (5.00 \text{ rad/s}^2)(25.0 \text{ s})$$

$$\omega_f = \boxed{125 \text{ rad/s}}$$

ADDITIONAL PRACTICE

1. In 1987, Takayuki Koike of Japan rode a unicycle nonstop for 160 km in less than 7 h. Suppose at some point in his trip Koike accelerated downhill. If the wheel's angular speed was initially 5.0 rad/s, what would the angular speed be after the wheel underwent an angular acceleration of 0.60 rad/s^2 for 0.50 min?

2. The moon orbits Earth in 27.3 days. Suppose a spacecraft leaves the moon and follows the same orbit as the moon. If the spacecraft has a constant angular acceleration of 1.0×10^{-10} rad/s^2, what is its angular speed after 12 h of flight? (Hint: At takeoff, the spaceship has the same angular speed as the moon.)

3. African baobab trees can have circumferences of up to 43 m. Imagine riding a bicycle around a tree this size. If, starting from rest, you travel a distance of 160 m around the tree with a constant angular acceleration of 5.0×10^{-2} rad/s^2, what will your final angular speed be?

4. In 1976, Kathy Wafler produced an unbroken apple peel 52.5 m in length. Suppose Wafler turned the apple with a constant angular acceleration of -3.2×10^{-5} rad/s^2, until her final angular speed was 0.080 rad/s. Assuming the apple was a sphere with a radius of 8.0 cm, calculate the apple's initial angular speed.

5. In 1987, a giant hanging basket of flowers with a mass of 4000 kg was constructed. The radius of the basket was 3.0 m. Suppose this basket was placed on the ground and an admiring spectator ran around it to see every detail again and again. At first the spectator's angular speed was 0.820 rad/s, but he steadily decreased his speed to 0.360 rad/s by the time he had traveled 20.0 m around the basket. Find the spectator's angular acceleration.

6. One of the largest scientific devices in the world is the particle accelerator at Fermilab, in Batavia, Illinois. The accelerator consists of a giant ring with a radius of 1.0 km. Suppose a maintenance engineer drives around the accelerator, starting at an angular speed of 5.0×10^{-3} rad/s and accelerating at a constant rate until one trip is completed in 14.0 min. Find the engineer's angular acceleration.

7. With a diameter of 207.0 m, the Superdome in New Orleans, Louisiana, is the largest "dome" in the world. Imagine a race held along the Superdome's outside perimeter. An athlete whose initial angular speed is 7.20×10^{-2} rad/s runs 12.6 rad in 4 min 22 s. What is the athlete's constant angular acceleration?

8. In 1986, Fred Markham rode a bicycle that was pulled by an automobile 200 m in 6.83 s. Suppose the angular speed of the bicycle's wheels increased steadily from 27.0 rad/s to 32.0 rad/s. Find the wheels' angular acceleration.

9. Consider a common analog clock. At midnight, the hour and minute hands coincide. Then the minute hand begins to rotate away from the hour hand. Suppose you adjust the clock by pushing the hour hand clockwise with a constant acceleration of 2.68×10^{-5} rad/s^2. What is the angular displacement of the hour hand after 120.0 s? (Note that the unaccelerated hour hand makes one full rotation in 12 h.)

10. Herman Bax of Canada grew a pumpkin with a circumference of 7 m. Suppose an ant crawled around the pumpkin's "equator." The ant started with an angular speed of 6.0×10^{-3} rad/s and accelerated steadily at a rate of 2.5×10^{-4} rad/s^2 until its angular speed was tripled. What was the ant's angular displacement?

11. Tal Burt of Israel rode a bicycle around the world in 77 days. If Burt could have ridden along the equator, his average angular speed would have been 9.0×10^{-7} rad/s. Now consider an object moving with this angular speed. How long would it take the object to reach an angular speed of 5.0×10^{-6} rad/s if its angular acceleration was 7.5×10^{-10} rad/s^2?

12. A coal-burning power plant in Kazakhstan has a chimney that is nearly 500 m tall. The radius of this chimney is 7.1 m at the base. Suppose a factory worker takes a 500.0 m run around the base of the chimney. If the worker starts with an angular speed of 0.40 rad/s and has an angular acceleration of 4.0×10^{-3} rad/s^2, how long will the run take?

Advanced Topics

Problem E

TANGENTIAL SPEED

PROBLEM

In about 45 min, Nicholas Mason inflated a weather balloon using only lung power. If a fly, moving with a tangential speed of 5.11 m/s, were to make exactly 8 revolutions around this inflated balloon in 12.0 s, what would the balloon's radius be?

SOLUTION

Given: $v_t = 5.11$ m/s

$\Delta\theta = 8$ rev

$\Delta t = 12.0$ s

Unknown: $\omega = ?$ $r = ?$

Use $\Delta\theta$ and Δt to calculate the fly's angular speed. Then rearrange the tangential speed equation to determine the balloon's radius.

$$\omega = \frac{\Delta\theta}{\Delta t} = \frac{(8 \text{ rev})\left(2\pi\frac{\text{rad}}{\text{rev}}\right)}{12.0 \text{ s}} = 4.19 \text{ rad/s}$$

$$v_t = r\omega$$

$$r = \frac{v_t}{\omega} = \frac{5.11 \text{ m/s}}{4.19 \text{ rad/s}} = \boxed{1.22 \text{ m}}$$

ADDITIONAL PRACTICE

1. The world's tallest columns, which stand in front of the Education Building in Albany, New York, are each 30 m tall. If a fly circles a column with an angular speed of 4.44 rad/s, and its tangential speed is 4.44 m/s, what is the radius of the column?

2. The longest dingoproof wire fence stretches across southeastern Australia. Suppose this fence were to have a circular shape. A rancher driving around the perimeter of the fence with a tangential speed of 16.0 m/s has an angular speed of 1.82×10^{-5} rad/s. What is the fence's radius and length (circumference)?

3. The smallest self-sustaining gas turbine has a tiny wheel that can rotate at 5.24×10^{3} rad/s. If the wheel rim's tangential speed is 131 m/s, what is the wheel's radius?

4. Earth's average tangential speed around the sun is about 29.7 km/s. If Earth's average orbital radius is 1.50×10^{8} km, what is its angular orbital speed in rad/s?

5. Two English engineers designed a ridable motorcycle that was less than 12 cm long. The front wheel's diameter was only 19.0 mm. Suppose this motorcycle was ridden so that the front wheel had an angular speed of 25.6 rad/s. What would the tangential speed of the front wheel's rim have been?

NAME ______________________ DATE ____________ CLASS ____________

Advanced Topics

Problem F

TANGENTIAL ACCELERATION

PROBLEM

In 1988, Stu Cohen made his kite perform 2911 figure eights in just 1 h. If the kite made a circular "loop" with a radius of 1.5 m and had a tangential acceleration of 0.6 m/s^2, what was the kite's angular acceleration?

SOLUTION

Given: $r = 1.5$ m $\qquad a_t = 0.6$ m/s^2

Unknown: $\alpha = ?$

Apply the tangential acceleration equation, solving for angular acceleration.

$$a_t = r\alpha$$

$$\alpha = \frac{a_t}{r} = \frac{0.6 \text{ m/s}^2}{1.5 \text{ m}} = \boxed{0.4 \text{ rad/s}^2}$$

ADDITIONAL PRACTICE

1. The world's largest aquarium, at the EPCOT center in Orlando, Florida, has a radius of 32 m. Bicycling around this aquarium with a tangential acceleration of 0.20 m/s^2, what would your angular acceleration be?
2. Dale Lyons and David Pettifer ran the London marathon in less than 4 h while bound together at one ankle and one wrist. Suppose they made a turn with a radius of 8.0 m. If their tangential acceleration was −1.44 m/s^2, what was their angular acceleration?
3. In 1991, Timothy Badyna of the United States ran 10 km backward in just over 45 min. Suppose Badyna made a turn, so that his angular speed decreased 2.4×10^{-2} rad/s in 6.0 s. If his tangential acceleration was −0.16 m/s^2, what was the radius of the turn?
4. Park Bong-tae of South Korea made 14 628 turns of a jump rope in 1.000 h. Suppose the rope's average angular speed was gained from rest during the first turn and that the rope during this time had a tangential acceleration of 33.0 m/s^2. What was the radius of the rope's circular path?
5. To "crack" a whip requires making its tip move at a supersonic speed. Kris King of Ohio achieved this with a whip 56.24 m long. If the tip of this whip moved in a circle and its angular speed increased from 6.00 rad/s to 6.30 rad/s in 0.60 s, what would the magnitude of the tip's tangential acceleration be?
6. In 1986 in Japan, a giant top with a radius of 1.3 m was spun. The top remained spinning for over an hour. Suppose these people accelerated the top from rest so that the first turn was completed in 1.8 s. What was the tangential acceleration of points on the top's rim?

Advanced Topics

Problem G

ROTATIONAL EQUILIBRIUM

PROBLEM

In 1960, a polar bear with a mass of 9.00×10^2 kg was discovered in Alaska. Suppose this bear crosses a 12.0 m long horizontal bridge that spans a gully. The bridge consists of a wide board that has a uniform mass of 2.50×10^2 kg and whose ends are loosely set on either side of the gully. When the bear is two-thirds of the way across the bridge, what is the normal force acting on the board at the end farthest from the bear?

SOLUTION

1. DEFINE **Given:**

m_b = mass of bridge = 2.50×10^2 kg

m_p = mass of polar bear = 9.00×10^2 kg

ℓ = length of bridge = 12.0 m

$g = 9.81 \text{ m/s}^2$

Unknown: $F_{n,1} = ?$

Diagram:

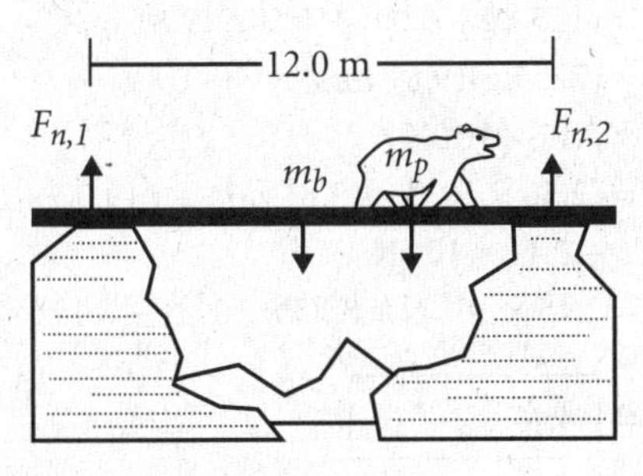

2. PLAN **Choose the equation(s) or situation:**

Apply the first condition of equilibrium: The unknowns in this problem are the normal forces exerted upward by the ground on either end of the board. The known quantities are the weights of the bridge and the polar bear. All of the forces are in the vertical (y) direction.

$$F_y = F_{n,1} + F_{n,2} - m_b g - m_p g = 0$$

Because there are two unknowns and only one equation, the solution cannot be obtained from the first condition of equilibrium alone.

Choose a point for calculating net torque: Choose the end of the bridge farthest from the bear as the pivot point. The torque produced by $F_{n,1}$ will be zero.

Apply the second condition of equilibrium: The torques produced by the bridge's and polar bear's weights are clockwise and therefore negative. The normal force on the end of the bridge opposite the axis of rotation exerts a counterclockwise (positive) torque.

$$\tau_{net} = -(m_b g)d_b - (m_p g)d_p + F_{n,2}d_2 = 0$$

The lever arm for the bridge's weight (d_b) is the distance from the bridge's center of mass to the pivot point, or half the bridge's length. The lever arm for the polar bear is two-thirds the bridge's length. The lever arm for the normal force farthest from the pivot equals the entire length of the bridge.

$$d_b = \tfrac{1}{2}\ell, \qquad d_p = \tfrac{2}{3}\ell, \qquad d_2 = \ell$$

The torque equation thus takes the following form:

$$\tau_{net} = -\frac{m_b g\ell}{2} - \frac{2m_p g\ell}{3} + F_{n,2}\ell = 0$$

Rearrange the equation(s) to isolate the unknowns:

$$F_{n,2} = \frac{m_b g\ell}{2\ell} + \frac{2m_p g\ell}{3\ell} = \left(\frac{m_b}{2} + \frac{2\,m_p}{3}\right)g$$

$$F_{n,1} = m_b g + m_p g - F_{n,2}$$

3. CALCULATE **Substitute the values into the equation(s) and solve:**

$$F_{n,2} = \left[\frac{2.50 \times 10^2 \text{ kg}}{2} + \frac{(2)(9.00 \times 10^2 \text{ kg})}{3}\right](9.81 \text{ m/s}^2)$$

$$F_{n,2} = (125 \text{ kg} + 6.00 \times 10^2 \text{ kg})(9.81 \text{ m/s}^2)$$

$$F_{n,2} = (725 \text{ kg})(9.81 \text{ m/s}^2)$$

$$F_{n,2} = 7.11 \times 10^3 \text{ N}$$

$$F_{n,1} = (2.50 \times 10^2 \text{ kg})(9.81 \text{ m/s}^2) + (9.00 \times 10^2 \text{ kg})(9.81 \text{ m/s}^2) - 7.11 \times 10^3 \text{ N}$$

$$F_{n,1} = 2.45 \times 10^3 \text{ N} + 8.83 \times 10^3 \text{ N} - 7.11 \times 10^3 \text{ N}$$

$$F_{n,1} = \boxed{4.17 \times 10^3 \text{ N}}$$

4. EVALULATE The sum of the upward normal forces exerted on the ends of the bridge must equal the weight of the polar bear and the bridge. (The individual normal forces change as the polar bear moves across the bridge.)

$$(4.17 \text{ kN} + 7.11 \text{ kN}) = (2.50 \times 10^2 \text{ kg} + 9.00 \times 10^2 \text{ kg})(9.81 \text{ m/s}^2)$$

$$11.28 \text{ kN} = 11.28 \times 10^3 \text{ N}$$

ADDITIONAL PRACTICE

1. The heaviest sea sponge ever collected had a mass of 40.0 kg, but after drying out, its mass decreased to 5.4 kg. Suppose two loads equal to the wet and dry masses of this giant sponge hang from the opposite ends of a horizontal meterstick of negligible mass and that a fulcrum is placed 70.0 cm from the larger of the two masses. How much extra force must be applied to the end of the meterstick with the smaller mass in order to provide equilibrium?

2. A Saguaro cactus with a height of 24 m and an estimated age of 150 years was discovered in 1978 in Arizona. Unfortunately, a storm toppled it in 1986. Suppose the storm produced a torque of 2.00×10^5 N•m that acted on the cactus. If the cactus could withstand a torque of only

1.2×10^5 N•m, what minimum force could have been applied to the cactus keep it standing? At what point and in what direction should this force have been applied? Assume that the cactus itself was very strong and that the roots were just pulled out of the ground.

3. In 1994, John Evans set a record for brick balancing by holding a load of bricks with a mass of 134 kg on his head for 10 s. Another, less extreme, method of balancing this load would be to use a lever. Suppose a board with a length of 7.00 m is placed on a fulcrum and the bricks are set on one end of the board at a distance of 2.00 m from the fulcrum. If a force is applied at a right angle to the other end of the board and the force has a direction that is 60.0° below the horizontal and away from the bricks, how great must this force be to keep the load in equilibrium? Assume the board has negligible mass.

4. In 1994, a vanilla ice lollipop with a mass of 8.8×10^3 kg was made in Poland. Suppose this ice lollipop was placed on the end of a lever 15 m in length. A fulcrum was placed 3.0 m from the lollipop so that the lever made an angle of 20.0° with the ground. If the force was applied perpendicular to the lever, what was the smallest magnitude this force could have and still lift the lollipop? Neglect the mass of the lever.

5. The Galápagos fur seals are very small. An average adult male has a mass of 64 kg, and a female has a mass of only 27 kg. Suppose one average adult male seal and one average adult female seal sit on opposite ends of a light board that has a length of 3.0 m. How far from the male seal should the board be pivoted in order for equilibrium to be maintained?

6. Goliath, a giant Galápagos tortoise living in Florida, has a mass of 3.6×10^2 kg. Suppose Goliath walks along a heavy board above a swimming pool. The board has a mass of 6.0×10^2 kg and a length of 15 m, and it is placed horizontally on the edge of the pool so that only 5.0 m of it extends over the water. How far out along this 5.0 m extension of the board can Goliath walk before he falls into the pool?

7. The largest pumpkin ever grown had a mass of 449 kg. Suppose this pumpkin was placed on a platform that was supported by two bases 5.0 m apart. If the left base exerted a normal force of 2.70×10^3 N on the platform, how far must the pumpkin have been from the platform's left edge? The platform had negligible mass.

8. In 1991, a giant stick of Brighton rock (a type of rock candy) was made in England. The candy had a mass of 414 kg and a length of 5.00 m. Imagine that the candy was balanced horizontally on a fulcrum. A child with a mass of 40.0 kg sat on one end of the stick. How far must the fulcrum have been from the child in order to maintain equilibrium?

Advanced Topics

Problem H

NEWTON'S SECOND LAW FOR ROTATION

PROBLEM

The giant sequoia General Sherman in California has a mass of about 2.00 × 10^6 kg, making it the most massive tree in the world. Its height of 83.0 m is also impressive. Imagine a uniform bar with the same mass and length as the tree. If this bar is rotated about an axis that is perpendicular to and passes through the bar's midpoint, how large an angular acceleration would result from a torque of 4.60 × 10^7 N·m? (Note: Assume the bar is thin.)

SOLUTION

Given: $M = 2.00 \times 10^6$ kg

$\ell = 83.0$ m

$\tau = 4.60 \times 10^7$ N•m

Unknown: $\alpha \times ?$

Calculate the bar's moment of inertia using the formula for a thin rod with the axis of rotation at its center.

$$I = \frac{1}{12}M\ell^2 = \frac{1}{12}(2.00 \times 10^6 \text{ kg})(83.0 \text{ m})^2 = 1.15 \times 10^9 \text{ kg•m}^2$$

Now use the equation for Newton's second law for rotating objects. Rearrange the equation to solve for angular acceleration.

$$\tau = I\alpha$$

$$\alpha = \frac{\tau}{I} = \frac{(4.60 \times 10^7 \text{ N•m})}{(1.15 \times 10^9 \text{ kg•m}^2)} = \boxed{4.00 \times 10^{-2} \text{ rad/s}^2}$$

ADDITIONAL PRACTICE

1. One of the largest Ferris wheels currently in existence is in Yokohama, Japan. The wheel has a radius of 50.0 m and a mass of 1.20 × 10^6 kg. If a torque of 1.0 × 10^9 N•m is needed to turn the wheel from a state of rest, what would the wheel's angular acceleration be? Treat the wheel as a thin hoop.

2. In 1992, Jacky Vranken from Belgium attained a speed of more than 250 km/h on just the back wheel of a motorcycle. Assume that all of the back wheel's mass is located at its outer edge. If the wheel has a mass of 22 kg and a radius of 0.36 m, what is the wheel's angular acceleration when a torque of 5.7 N•m acts on the wheel?

3. In 1995, a fully functional pencil with a mass of 24 kg and a length of 2.74 m was made. Suppose this pencil is suspended at its midpoint and a force of 1.8 N is applied perpendicular to its end, causing it to rotate. What is the angular acceleration of the pencil?

4. The turbines at the Grand Coulee Third Power Plant in the state of Washington have rotors with a mass of 4.07×10^5 kg and a radius of 5.0 m each. What angular acceleration would one of these rotors have if a torque of 5.0×10^4 N•m were applied? Assume the rotor is a uniform disk.

5. J. C. Payne of Texas amassed a ball of string that had a radius of 2.00 m. Suppose a force of 208 N was applied tangentially to the ball's surface in order to give the ball an angular acceleration of 3.20×10^{-2} rad/s^2. What was the ball's moment of inertia?

6. The heaviest member of British Parliament ever was Sir Cyril Smith. Calculate his peak mass by finding first his moment of inertia from the following situation. If Sir Cyril were to have ridden on a merry-go-round with a radius of 8.0 m, a torque of 7.3×10^3 N•m would have been needed to provide him with an angular acceleration of 0.60 rad/s^2.

7. In 1975, a centrifuge at a research center in England made a carbon-fiber rod spin about its center so fast that the tangential speed of the rod's tips was about 2.0 km/s. The length of the rod was 15.0 cm. If it took 80.0 s for a torque of 0.20 N•m to bring the rod to rest from its maximum speed, what was the rod's moment of inertia?

8. The largest tricycle ever built had rear wheels that were almost 1.70 m in diameter. Neglecting the mass of the spokes, the moment of inertia of one of these wheels is equal to that of a thin hoop rotated about its symmetry axis. Find the wheel's moment of inertia and its mass if a torque of 125 N•m is applied to the wheel so that in 2.0 s the wheel's angular speed increases from 0 rad/s to 12 rad/s.

9. In 1990, a cherry pie with a radius of 3.00 m and a mass of 17×10^3 kg was baked in Canada. Suppose the pie was placed on a light rotating platform attached to a motor. If this motor brought the angular speed of the pie from 0 rad/s to 3.46 rad/s in 12 s, what was the torque the motor must have produced? Assume the mass of the platform was negligible and the pie was a uniformly solid disk.

10. In just over a month in 1962, a shaft almost 4.00×10^8 m deep and with a radius of 4.0 m was drilled in South Africa. The mass of the soil taken out was about 1.0×10^8 kg. Imagine a rigid cylinder with a mass, radius, and length equal to these values. If this cylinder rotates about its symmetry axis so that it undergoes a constant angular acceleration from 0 rad/s to 0.080 rad/s in 60.0 s, how large a torque must act on the cylinder?

11. In 1993, a bowl in Canada was filled with strawberries. The mass of the bowl and strawberries combined was 2390 kg, and the moment of inertia about the symmetry axis was estimated to be 2.40×10^3 kg•m^2. Suppose a constant angular acceleration was applied to the bowl so that it made its first two complete rotations in 6.00 s. How large was the torque that acted on the bowl?

12. A steel ax with a mass of 7.0×10^3 kg and a length of 18.3 m was made in Canada. If Paul Bunyan were to take a swing with such an ax, what torque would he have to produce in order for the blade to have a tangential acceleration of 25 m/s^2? Assume that the blade follows a circle with a radius equal to the ax handle's length and that nearly all of the mass is concentrated in the blade.

NAME ______________________ DATE ______________ CLASS ______________

Advanced Topics

Problem I

CONSERVATION OF ANGULAR MOMENTUM

PROBLEM

The average distance from Earth to the moon is 3.84 × 10^5 km. The average orbital speed of the moon when it is at its average distance from Earth is 3.68 × 10^3 km/h. However, in 1912 the average orbital speed was 3.97 × 10^3 km/h, and in 1984 it was 3.47 × 10^3 km/h. Calculate the distances that correspond to the 1912 and 1984 orbital speeds, respectively.

SOLUTION

1. DEFINE

Given:

$r_{avg} = 3.84 \times 10^5$ km

$v_{avg} = 3.68 \times 10^3$ km/h

$v_1 = 3.97 \times 10^3$ km/h

$v_2 = 3.47 \times 10^3$ km/h

Unknown: $r_1 = ?$ $r_2 = ?$

2. PLAN

Choose the equation(s) or situation: Because there are no external torques, the angular momentum of the Earth-moon system is conserved.

$$L_{avg} = L_1 = L_2$$

$$I_{avg}\,\omega_{avg} = I_1\,\omega_1 = I_2\,\omega_2$$

If the moon is treated as a point mass revolving around a central axis, its moment of inertia is simply mr^2, and the conservation of momentum expression takes the following form:

$$m_{moon}\,(v_{avg})^2\left(\frac{v_{avg}}{r_{avg}}\right) = m_{moon}\,(r_1)^2\left(\frac{v_1}{r_1}\right) = m_{moon}\,(r_2)^2\left(\frac{v_2}{r_2}\right)$$

Because the mass of the moon is unchanged, the mass term cancels, and the equation reduces to the following:

$$r_{avg}\,v_{avg} = r_1\,v_1 = r_2\,v_2$$

Rearrange the equation(s) to isolate the unknown(s):

$$r_1 = \frac{r_{avg}\,v_{avg}}{v_1} \qquad r_2 = \frac{r_{avg}v_{avg}}{v_2}$$

3. CALCULATE

Substitute the values into the equation(s) and solve:

$$r_1 = \frac{(3.84 \times 10^5\ \text{km})(3.68 \times 10^3\ \text{km/h})}{(3.97 \times 10^3\ \text{km/h})} = \boxed{3.56 \times 10^5\ \text{km}}$$

$$r_2 = \frac{(3.84 \times 10^5\ \text{km})(3.68 \times 10^3\ \text{km/h})}{(3.47 \times 10^3\ \text{km/h})} = \boxed{4.07 \times 10^5\ \text{km}}$$

4. EVALUATE

Because angular momentum is conserved in the absence of external torques, the tangential orbital speed of the moon is greater than its average value when the moon is closer to Earth. Similarly, the smaller tangential orbital speed occurs when the moon is farther from Earth.

ADDITIONAL PRACTICE

1. Encke's comet revolves around the sun in a period of just over three years (the shortest period of any comet). The closest it approaches the sun is 4.95×10^7 km, at which time its orbital speed is 2.54×10^5 km/h. At what distance from the sun would Encke's comet have a speed equal to 1.81×10^5 km/h?

2. In 1981, Sammy Miller reached a speed of 399 km/h on a rocket-powered ice sled. Suppose the sled, moving at its maximum speed, was hooked to a post with a radius of 0.20 m by a light cord with an unknown initial length. The rocket engine was then turned off, and the sled began to circle the post with negligible resistance as the cord wrapped around the post. If the speed of the sled after 20 turns was 456 km/h, what was the length of the unwound cord?

3. Earth is not a perfect sphere, in part because of its rotation about its axis. A point on the equator is in fact over 21 km farther from Earth's center than is the North pole. Suppose you model Earth as a solid clay sphere with a mass of 25.0 kg and a radius of 15.0 cm. If you begin rotating the sphere with a constant angular speed of 4.70×10^{-3} rad/s (about the same as Earth's), and the sphere continues to rotate without the application of any external torques, what will the change in the sphere's moment of inertia be when the final angular speed equals 4.74×10^{-3} rad/s?

4. In 1971, a model plane built in the Soviet Union by Leonid Lipinsky reached a speed of 395 km/h. The plane was held in a circular path by a control line. Suppose the plane ran out of gas while moving at its maximum speed and Lipinsky pulled the line in to bring the plane home while it continued in a circular path. If the line's initial length is 1.20×10^2 m and Lipinsky shortened the line by 0.80 m every second, what was the plane's speed after 32 s?

5. The longest spacewalk by a team of astronauts lasted more than 8 h. It was performed in 1992 by three crew members from the space shuttle *Endeavour.* Suppose that during the walk two astronauts with equal masses held the opposite ends of a rope that was 10.0 m long. From the point of view of the third astronaut, the other two astronauts rotated about the midpoint of the rope with an angular speed of 1.26 rad/s. If the astronauts shortened the rope equally from both ends, what was their angular speed when the rope was 4.00 m long?

6. In a problem in the previous section, a cherry pie with a radius of 3.00 m and a mass of 17×10^3 kg was rotated on a light platform. Suppose that when the pie reached an angular speed of 3.46 rad/s there was no net torque acting on it. Over time, the filling in the pie began to move outward, changing the pie's moment of inertia. Assume the pie acted like a uniform, rigid, spinning disk with a mass of 16.80×10^3 kg combined with a 0.20×10^3 kg particle. If the smaller mass shifted from a position 2.50 m from the center to one that was 3.00 m from the center, what was the change in the angular speed of the pie?

Advanced Topics

Problem J

CONSERVATION OF MECHANICAL ENERGY

PROBLEM

In 1990, Eddy McDonald of Canada completed 8437 loops with a yo-yo in an hour, setting a world record. Assume that the yo-yo McDonald used had a mass of 6.00×10^{-2} kg. The yo-yo descended from a height of 0.600 m down a vertical string and had a linear speed of 1.80 m/s by the time it reached the bottom of the string. If its final angular speed was 82.6 rad/s, what was the yo-yo's moment of inertia?

SOLUTION

1. DEFINE **Given:** $m = 6.00 \times 10^{-2}$ kg $\quad h = 0.600$ m

$v_f = 1.80$ m/s $\quad \omega_f = 82.6$ rad/s

$g = 9.81$ m/s^2

Unknown: $I = ?$

2. PLAN **Choose the equation(s) or situation:** Apply the principle of conservation of mechanical energy.

$$ME_i = ME_f$$

Initially, the system possesses only gravitational potential energy.

$$ME_i = PE_g = mgh$$

When the yo-yo reaches the bottom of the string, this potential energy has been converted to translational and rotational kinetic energy.

$$ME_f = KE_{trans} + KE_{rot} = \tfrac{1}{2}mv_f^2 + \tfrac{1}{2}I\omega_f^2$$

Equate the initial and final mechanical energy.

$$mgh = \tfrac{1}{2}mv_f^2 + \tfrac{1}{2}I\omega_f^2$$

Rearrange the equation(s) to isolate the unknown(s):

$$\tfrac{1}{2}I\omega_f^2 = mgh - \tfrac{1}{2}mv_f^2 \qquad I = \frac{2\,mgh - mv_f^2}{\omega_f^2} = \frac{m(2gh - v_f^2)}{\omega_f^2}$$

3. CALCULATE **Substitute the values into the equation(s) and solve:**

$$I = \frac{(6.00 \times 10^{-2}\ \text{kg})[(2)\ (9.81\ \text{m/s}^2)(0.600\ \text{m}) - (1.80\ \text{m/s})^2]}{(82.6\ \text{rad/s})^2}$$

$$I = \frac{(6.00 \times 10^{-2}\ \text{kg})(8.6\ \text{m}^2/\text{s}^2)}{(82.6\ \text{rad/s})^2} = \boxed{7.6 \times 10^{-5}\ \text{kg}\bullet\text{m}^2}$$

4. EVALUATE Assuming that the yo-yo can be approximated by a solid disk with a central axis of rotation, the yo-yo's moment of inertia is described by the equation $I = \tfrac{1}{2}MR^2$. From the calculated value for I, the yo-yo's radius can be found to be 5.0 cm.

$$R = \sqrt{\frac{2I}{M}} = \sqrt{\frac{(2)(7.6 \times 10^{-5}\ \text{kg}\bullet\text{m}^2)}{6.00 \times 10^{-2}\ \text{kg}}} = 0.050\ \text{m}$$

NAME ______________________ DATE ______________ CLASS ______________

ADDITIONAL PRACTICE

1. In 1993, a group of students in England made a giant yo-yo that was 10 ft in diameter and had a mass of 407 kg. With the use of a crane, the yo-yo was launched from a height of 57.0 m above the ground. Suppose the linear speed of the yo-yo at the end of its descent was 12.4 m/s. If the angular speed at the end of the descent was 28.0 rad/s, what was the yo-yo's moment of inertia?

2. In 1994, a bottle over 3 m in height and with a radius at its base of 0.56 m was made in Australia. Treat the bottle as a thin-walled cylinder rotating about its symmetry axis, which has the same rotational properties as a thin hoop rotating about its symmetry axis. What is the linear speed that the bottle acquires after rolling down a slope with a height of 5.0 m? Do you need to know the mass of the bottle?

3. In 1988, a cheese with a mass of 1.82×10^4 kg was made in Wisconsin. Suppose the cheese had a cylindrical shape. The cheese was set to roll along a horizontal road with an undetermined speed. The road then went uphill, and the cheese rolled up the hill until its vertical displacement was 1.2 m, at which point it came to a stop. Assuming that there was no slipping between the rim of the cheese and the ground, calculate the initial linear speed of the cheese.

4. In 1982, a team of ten people rolled a cylindrical barrel with a mass of 64 kg for almost 250 km without stopping. Imagine that at some point during the trip the barrel was stopped at the crest of a steep hill. The barrel was accidentally released and rolled down the hill. If the linear speed of the barrel at the bottom of the slope was 12.0 m/s, how high was the hill? Assume that the barrel's moment of inertia was equal to $0.80\ mr^2$.

5. A potato with a record-breaking mass of 3.5 kg was grown in 1994. Suppose a child saw this potato and decided to pretend it was a soccer ball. The child kicked the potato so that it rolled without slipping at a speed of 5.4 m/s. The potato rolled up a slope with a 30.0° incline. Assuming that the potato could be approximated as a uniform, solid sphere with a radius of 7.0 cm, what was the distance along the slope that the potato rolled before coming to a stop?

6. In 1992, an artificial egg with a mass of 4.8×10^3 kg was made in Australia. Assume that the egg is a solid sphere with a radius of 2.0 m. Calculate the minimum height of a slope that the egg rolls down if it is to reach an angular speed of 5.0 rad/s at the bottom of the slope. What is the translational kinetic energy of the egg at the bottom of the slope?

7. An onion grown in 1994 had a record-breaking mass of 5.55 kg. Assume that this onion can be approximated by a uniform, solid sphere. Suppose the onion rolled down an inclined ramp that had a height of 1.40 m. What was the onion's rotational kinetic energy? Assume that there was no slippage between the ramp and the onion's surface.

NAME ______________________ DATE ____________ CLASS ______________

Advanced Topics

Problem K

BERNOULLI'S EQUATION

PROBLEM

The widest road tunnel in the world is located in California. The tunnel has a cross-sectional area of about 4.00×10^2 m^2. On the other hand, the Three Rivers water tunnel in Georgia has a cross-sectional area of only 8.0 m^2. Imagine connecting together two tunnels with areas equal to these along a flat region and setting fresh water to flow at 4.0 m/s in the narrower tunnel. If the pressure in the wider tunnel is 1.10×10^5 Pa, what is the pressure in the narrower tunnel?

SOLUTION

1. DEFINE

Given: $A_1 = 8.0 \text{ m}^2$ $\quad A_2 = 4.00 \times 10^2 \text{ m}^2$

$\rho = 1.00 \times 10^3 \text{ kg/m}^3$

$v_1 = 4.0 \text{ m/s}$

$P_2 = 1.10 \times 10^5 \text{ Pa}$

Unknown: $P_1 = ?$

2. PLAN

Choose the equation(s) or situation: Because this problem involves fluid flow, it requires the application of Bernoulli's equation.

$$P_1 + \tfrac{1}{2}\rho v_1^2 + \rho g h_1 = P_2 + \tfrac{1}{2}\rho v_2^2 + \rho g h_2$$

The flow of water is horizontal, so h_1 and h_2 are equal.

$$P_1 + \tfrac{1}{2}\rho v_1^2 = P_2 + \tfrac{1}{2}\rho v_2^2$$

To find the speed of the flowing water in the wider tunnel, use the continuity equation.

$$A_1 v_1 = A_2 v_2$$

Substitute this equation for v_2 into Bernoulli's equation.

$$P_1 + \tfrac{1}{2}\rho v_1^2 = P_2 + \tfrac{1}{2}\rho\left(\frac{A_1}{A_2}v_1\right)^2$$

Rearrange the equation(s) to isolate the unknown(s):

$$P_1 = P_2 + \tfrac{1}{2}\rho v_1^2\left[\left(\frac{A_1}{A_2}\right)^2 - 1\right]$$

3. CALCULATE

Substitute the values into the equation(s) and solve:

$$P_1 = 1.10 \times 10^5 \text{ Pa} + \tfrac{1}{2}(1.00 \times 10^3 \text{ kg/m}^3)(4.0 \text{ m/s})^2\left[\left(\frac{8.0 \text{ m}^2}{4.00 \times 10^2 \text{ m}^2}\right)^2 - 1\right]$$

$$P_1 = 1.10 \times 10^5 \text{ Pa} + (8.0 \times 10^3 \text{ N}\bullet\text{m/m}^3)(4.0 \times 10^{-4} - 1)$$

$$P_1 = 1.10 \times 10^5 \text{ Pa} - (8.0 \times 10^3 \text{ Pa})(0.9996)$$

$$P_1 = 1.10 \times 10^5 \text{ Pa} - 8.0 \times 10^3 \text{ Pa}$$

$$P_1 = \boxed{1.02 \times 10^5 \text{ Pa}}$$

NAME ______________________ DATE ______________ CLASS ______________

4. EVALUATE The pressure of the water increases when it flows into the wider tunnel. The speed of the water's flow decreases, as indicated by the continuity equation.

$$v_2 = (4.0 \text{ m/s})(8/400) = 8 \times 10^{-2} \text{ m/s}$$

ADDITIONAL PRACTICE

1. The Chicago system of sewer tunnels has a total length of about 200 km. The tunnels are all at the same level and vary in diameter from less than 3 m to 10 m. Consider a connection between a wide tunnel and a narrow tunnel. The pressure in the wide tunnel is 12 percent greater than the pressure in the narrow one, and the speeds of flowing water in the wide and the narrow tunnels are 0.60 m/s and 4.80 m/s, respectively. Based on this information, find the pressure in the wide tunnel.

2. A certain New York City water-supply tunnel is almost 170 km long, is all at the same level, and has a diameter of 4.10 m. Suppose water flows through the tunnel at a speed of 3.0 m/s until it reaches a narrow section where the tunnel's diameter is 2.70 m. The pressure in the narrow section is 82 kPa. Use the continuity equation to find the water's speed in the narrow section of the tunnel. Then find the pressure in the wide portion of the tunnel.

3. An enormous open vat owned by a British cider company has a volume of nearly 7000 m^3. If a small hole is drilled halfway down the side of this vat when it is full of cider, the cider will hit the ground 19.7 m away from the bottom of the vat. How tall is the vat?

4. The tallest cooling tower in the world is in Germany. Consider a water pipe coming down from the top of the tower. The pipe is punctured near the bottom, which causes the water to flow from the puncture hole with a speed of 59 m/s. How high is the tower? Assume the pressure inside the pipe is equal to the pressure outside the pipe.

5. A water tower built in Oklahoma has a capacity of 1893 m^3 and is about 66.0 m high. Suppose a little hole with a cross-sectional area of 10.0 cm^2 is drilled near the bottom of the tower's water tank. At what speed does the water initially flow through the hole? Assume that the air pressure inside and outside the tank is the same.

6. The Nurek Dam, in Tajikistan, is the tallest dam in the world—its height is more than 300 m. Suppose the difference in water levels on the opposite sides of the dam is 3.00×10^2 m. If a small crack appears in the dam near the lower water level, at what speed will the water stream leave the crack? Assume the air pressure on either side of the dam is the same.

7. The world's largest litter bin has a volume of more than 40 m^3 and is 6.0 m tall. If this bin is filled with water and then a hole with an area of 0.16 cm^2 is drilled near the bottom, at what speed will the water leave the bin? Assume that the water level in the bin drops slowly and that the bin is open to the air.

NAME ______________________ DATE ____________ CLASS ____________

Advanced Topics

Problem L

THE IDEAL GAS LAW

PROBLEM

A hot-air balloon named *Double Eagle V* traveled a record distance of more than 8000 km from Japan to California in 1981. The volume of the balloon was 1.13×10^4 m^3. If the balloon contained 2.84×10^{29} gas particles that had an average temperature of 355 K, what was the absolute pressure of the gas in the balloon?

SOLUTION

1. DEFINE **Given:**

$V = 1.13 \times 10^4 \text{ m}^3$

$N = 2.84 \times 10^{29}$ particles

$T = 355$ K

$k_B = 1.38 \times 10^{-23}$ J/K

Unknown: $P = ?$

2. PLAN **Choose the equation(s) or situation:** To find the pressure of the gas, use the ideal gas law.

$$PV = Nk_BT$$

Rearrange the equation(s) to isolate the unknown(s):

$$P = \frac{Nk_BT}{V}$$

3. CALCULATE **Substitute the values into the equation(s) and solve:**

$$P = \frac{(2.84 \times 10^{29} \text{ particles})(1.38 \times 10^{-23} \text{ J/K})(355 \text{ K})}{1.13 \times 10^4 \text{ m}^3}$$

$$P = \boxed{1.23 \times 10^5 \text{ Pa}}$$

4. EVALUATE The pressure inside the balloon is about 20 percent greater than standard air pressure. This pressure corresponds to the higher temperature of the air in the balloon. The hot air's temperature is also nearly 20 percent greater than 300 K.

ADDITIONAL PRACTICE

1. The official altitude record for a balloon was set in 1961 by two American officers piloting a high-altitude helium balloon with a volume of 3.4 $\times 10^5$ m^3. Assume that the temperature of the gas was 280 K. If the balloon contained 1.4×10^{30} atoms of helium, find the absolute pressure in the balloon at the maximum altitude of 35 km.

2. The estimated number of locusts that made up a swarm that infested Nebraska in 1874 was 1.2×10^{13}. This number was about 7000 times the total human population of Earth back then and about 2000 times the total human population today. It is, however, only a few billionths of the number of molecules in a liter of gas. If a container with a volume of 1.0×10^{-3} m^3

is filled with 1.2×10^{13} gas molecules maintained at a temperature of 300.0 K, what is the pressure of the gas in the container?

3. In terms of volume, the largest pyramid in the world is not in Egypt but in Mexico. The Pyramid of the Sun at Teotihuacan has a volume of 3.3×10^6 m^3. If you fill a balloon to this same volume with 1.5×10^{32} molecules of nitrogen at a temperature of 360 K, what will the absolute pressure of the gas be?

4. A snow palace more than 30 m high was built in Japan in 1994. Suppose a container with the same volume as this snow palace is filled with 1.00×10^{27} molecules. If the temperature of the gas is 2.70×10^2 K and the gas pressure is 36.2 Pa, what is the volume of the gas?

5. Suppose the volume of a balloon decreases so that the temperature of the balloon decreases from 280 K to 240 K and its pressure drops from 1.6×10^4 Pa to 1.7×10^4 Pa. What is the new volume of the gas?

6. The longest navigable tunnel in the world was built in France. Suppose the entire tunnel, which has a cross-sectional area of 2.50×10^2 m^2, is filled with air at a temperature of 3.00×10^2 K and a pressure of 101 kPa. If the tunnel contains 4.34×10^{31} molecules, what is the volume and length of the tunnel?

7. A balloon is filled with 7.36×10^4 m^3 of hot air. If the pressure inside the balloon is 1.00×10^5 Pa and there are 1.63×10^{30} particles of air inside, what is the average temperature of the air inside the balloon?

8. In 1993, a group of American researchers drilled a 3053 m shaft in the ice sheet of Greenland. Suppose the cross-sectional area of the shaft is 0.040 m^2. If the air in the shaft consists of 3.6×10^{27} molecules at an average pressure of 105 kPa, what is the air's average temperature?

9. The cylinder of the largest steam engine had a radius of 1.82 m. Suppose the length of the cylinder is six times the radius. Steam at a pressure of 2.50×10^6 Pa and a temperature of 495 K enters the cylinder when the piston has reduced the volume in the cylinder to 3.00 m^3. The piston is then pushed outward until the volume of the steam in the cylinder is 57.0 m^3. If the pressure of the steam after expansion is 1.01×10^5 Pa, what is the temperature of the steam?

NAME ______________________ DATE ______________ CLASS ______________

Advanced Topics

Problem M

INDUCTION IN GENERATORS

PROBLEM

In Virginia, a reversible turbine was built to serve as both a power generator and a pump. When operating as a generator, the turbine can deliver almost 5×10^8 W of power. The normal angular speed of the turbine rotor is 26.9 rad/s. Suppose the generator produces a maximum emf of 12 kV at this angular speed. If the generator's magnetic field strength is 0.12 T, how many turns of wire, each with an area of 40.0 m^2, are used in the generator?

SOLUTION

1. DEFINE **Given:**

$\omega = 26.9$ rad/s

$B = 0.12$ T

$A = 40.0\ m^2$

maximum emf $= 12$ kV $= 12 \times 10^3$ V

Unknown: $N = ?$

2. PLAN **Choose the equation(s) or situation:** Use the maximum emf equation for a generator.

$$\text{maximum emf} = NBA\omega$$

Rearrange the equation(s) to isolate the unknown(s):

$$N = \frac{\text{maximum emf}}{AB\omega}$$

3. CALCULATE **Substitute values into the equation(s) and solve:**

$$N = \frac{(12 \times 10^3\ \text{V})}{(40.0\ \text{m}^2)(0.12\ \text{T})(26.9\ \text{rad/s})} = \boxed{93\ \text{turns}}$$

4. EVALUATE Although the magnetic field of the generator is fairly large and the area of the loops is very large, several turns of wire are needed to produce a large maximum emf. This gives some indication that at least one—and often several—of the physical attributes of a generator must be large to induce a sizable emf.

ADDITIONAL PRACTICE

1. A gas turbine with rotors that are 5.0 cm in diameter was built in 1989. The turbine's rotors spin with a frequency of 833 Hz. Suppose a coil of wire has a cross-sectional area equal to that of the turbine rotor. This coil turns with the same frequency as the turbine rotor and is perpendicular to a 8.0×10^{-2} T magnetic field. If the maximum emf induced in the coil is 330 V, how many turns of wire are there in the coil?

2. A company in Michigan makes solar-powered lawn mowers. The batteries can run a 3 kW motor for over an hour, and the blades spin at 335 rad/s, which is about 30 percent faster than the blades on a gasoline-powered mower. Suppose the kinetic energy of rotation of the blades is used to generate electricity. The generator is built so that the coil turns at 335 rad/s, creating a maximum emf of 214 V. The coil is placed in a 8.00×10^{-2} T magnetic field. Each of the turns of wire on the rotating coil has an area of 0.400 m^2. Use this information to calculate the number of turns of wire.

3. Tom Archer designed a self-propelled vertical Catherine wheel (its rim is used to mount fireworks) that was 19.3 m in diameter. In 1994, the wheel was successfully demonstrated. It made a few turns at an average angular speed of 0.52 rad/s. If a similar wheel with exactly 40 turns of wire wrapped around the rim is placed in a uniform magnetic field and rotated about an axis that is along the wheel's diameter, an emf will be generated. Suppose the maximum induced emf is 2.5 V. If the angular speed of the wheel's rotation is 0.52 rad/s, what is the magnitude of the magnetic field strength?

4. The Garuda, an airplane propeller designed in Germany in 1919, was 6.90 m across and had an angular speed of 57.1 rad/s during flight. Consider a generator producing a maximum emf of 8.00×10^3 V. If the rotor has 236 square turns, each with 6.90 m sides, and if the angular speed of rotation equals 57.1 rad/s, what is the magnitude of the magnetic field in which the rotor must be turned?

5. In Japan, a 5 mm long working model of a car has been built. The motor, less than 1 mm long, uses a coil with 1000 turns of wire and is powered by a 3.0 V emf source. Consider a mini-generator that uses a coil with exactly 1000 turns of wire, each with an area of 8.0 cm^2. If the coil is placed in a 2.4×10^{-3} T magnetic field, what is the angular speed in rad/s needed to produce a maximum emf of 3.0 V?

6. In 1995, a turbine was built that had a rotor shaft suspended magnetically, almost fully eliminating friction. This allows the extremely high angular speeds needed to create a large emf to be achieved. Suppose the turbine turns a coil that contains exactly 640 turns of wire, each with an area of 0.127 m^2. This generator produces a maximum emf of 24.6 kV while rotating in a 8.00×10^{-2} T magnetic field. What is the angular speed in rad/s of the coil?

7. At the University of Virginia is a centrifuge whose rotor is magnetically supported in a vacuum; this allows for extremely low retarding forces. The frequency of rotation of this centrifuge is 1.0×10^3 Hz. Consider an electrical generator with this same frequency. The coil is placed in a magnetic field that has a magnitude of 0.22 T. If the coil of the generator has 250 circular loops, each with a 12 cm radius, what is the maximum emf that can be induced?